Mechanics
MOLECULAR PHYSICS AND HEAT

AF808380

MAXWELL PRESS

Mechanics
MOLECULAR PHYSICS AND HEAT

ROBERT ANDREWS MILLIKAN, PH.D.
Winner of Nobel Prize for Physics
Assistant Professor of Physics in the University of Chicago

MAXWELL PRESS
Chennai Trichy NewDelhi

MAXWELL PRESS

ISBN 978-93-90063-48-2 Maxwell Press

No. 44, Nallathambi Street,
Printed and bound in India Triplicane, Chennai 600 005

MJP 894

Publisher : C. Janarthanan

Publisher's Note

The legacy of a country is in its varied cultural heritage, historical literature, developments in the field of economy and science. The top nations in the world are competing in the field of science, economy and literature. This vast legacy has to be conserved and documented so that it can be bestowed to the future generation. The knowledge of this legacy is slowly getting perished in the present generation due to lack of documentation.

Keeping this in mind, the concern with retrospective acquiring of rare books has been accented recently by the burgeoning reprint industry. Maxwell Press is gratified to retrieve the rare collections with a view to bring back those books that were landmarks in their time.

In this effort, a series of rare books would be republished under the banner, "Maxwell Press". The books in the reprint series have been carefully selected for their contemporary usefulness as well as their historical importance within the intellectual. We reconstruct the book with slight enhancements made for better presentation, without affecting the contents of the original edition.

Most of the works selected for republishing covers a huge range of subjects, from history to anthropology. We believe this reprint edition will be a service to the numerous researchers and practitioners active in this fascinating field. We allow readers to experience the wonder of peering into a scholarly work of the highest order and seminal significance.

Maxwell Press

PREFACE

This book is neither a laboratory manual in the ordinary sense of the term, nor yet is it simply a class-room text. It is intended to take the place of both. It represents the first portion of a college course in General Physics in which the primary object has been to establish an immediate and vital connection between theory and experiment. Of course such connection always exists in the mind of the teacher; but the use in class-room and laboratory of separate texts, separate courses, and separate instructors is on the whole unfavorable to making it clear to the student. The student who takes an experimental course which is out of immediate connection with class-room discussion, who is provided in the laboratory with an isolated set of directions, or with a laboratory manual which is essentially a compendium of directions for all conceivable experiments, may perhaps in some cases obtain, with the aid of references to text-books, a comprehensive grasp of the theory and bearings of his experiment; but it is safe to say that in a great majority of cases he does not do so. The most serious criticism which can be urged against modern laboratory work in Physics is that it often degenerates into a servile following of directions, and thus loses all save a purely manipulative value. Important as is dexterity in the handling and adjustment of apparatus, it can not be too strongly emphasized that it is *grasp of principles*, not *skill in manipulation* which should be the primary object of General Physics courses.

Furthermore, an intimate connection between lecture and laboratory work is no less important from the standpoint of the former than of the latter. Without the fixing power of laboratory applications, a thorough grasp of physical principles is seldom, or never, gained. This is particularly true in Mechanics, the most fundamental of all the branches of Physics, for it is only through it that the door is opened to insight into the theories of Heat, Sound, Light, and Electricity.

In the second place, this book represents an attempt to teach thoroughly a *few fundamental principles* rather than to present superficially a large mass of facts. Most of the general texts which combine a full presentation of facts, with a satisfactory discussion of their relation to theory, have grown too bulky for general class-room use. On the other hand, other texts have appeared in which the necessity for condensation has resulted in such abridgment of discussion that there is little left but a skeleton of experimental and theoretical *results*.

In attempting to avoid both of these extremes, a selection has been made for the present course of such principles only as can be most effectively presented in connection with laboratory demonstrations. This course is presented in the first third (twelve weeks) of a year of General Physics in the Junior College at the University of Chicago. The time is divided nearly equally between class-room and laboratory work; but the former is wholly occupied with the discussion and application to practical problems, of the twenty-three principles presented in the text. No demonstration lectures whatever are given. The second third of the college year is occupied with the presentation, by the same general plan, of those parts of Electricity, Light, and Sound which can be most profitably studied in connection with laboratory instruments and methods. This course is embodied in a second volume. The last third of the year is devoted wholly to demonstration lectures upon subjects which have been omitted from the preceding courses because they are more suitable to lecture than to laboratory methods of presentation. Such are, for example, Static Electricity, Electric Radiation, the Discharge of Electricity through Gases, the Radiation, Absorption, Polarization, and Interference of Light, Physiological Optics and Acoustics; in a word, all phenomena the presentation of which requires primarily qualitative rather than quantitative experiment. This demonstration lecture course is given last instead of first because a thorough grounding in the fundamental principles of physical measurement is deemed absolutely essential to the intelligent following of lectures of college grade in any branch of Physics.

A third aim has been to prepare a book which should represent a continuous *course* in the study of the principles of experimental Physics, rather than a *compendium* either of facts or experiments.

Hence, in the first place, all purely manipulative experiments have been altogether omitted. A student is not required to make a useless measurement for the sake of learning to use an instrument; he rather learns to use the instrument for the sake of making a needed measurement. The inversion of this order has invariably weakened interest in laboratory work. In the second place, all needless repetition of slight variations of the same experiment has been avoided. For example, in the subject of Heat, there is but one general principle involved in the method of mixtures, whether it be applied to the determination of the specific heat of a solid, or of a liquid, the latent heat of fusion, or of vaporization. Hence it has been illustrated in this course by but one laboratory exercise. Similarly, the usual half-dozen or more experiments upon the densities of liquids and solids have been reduced to two. In a word, experiments have been made incidental to the study of principles, not principles incidental to the study of experiments.

Fourthly, an especial effort has been made *to present Physics as a science of exact measurement.* Much harm is often done by attempting to teach a course in an exact science with the aid of instruments capable of giving results which can be called quantitative by courtesy only, and which therefore foster the impression that science is after all very inexact. The apparatus which is used in this course has therefore been selected and designed with a special reference to its ability to yield accurate results in the hands of average students. All of the new pieces, such as the acceleration machine (p. 11), the model balance (p. 36), the ballistie pendulums (pp. 54 and 62), the Young's modulus (p. 67), the torsion machines (p. 75), the inertia disc (p. 82), the pendulum arrangement (p. 97), the centripetal machine (p. 102), the pressure apparatus (p. 117), the form of air thermometer (p. 131), the vapor pressure arrangement (p. 158), have been designed in this laboratory, slight modifications having in some instances been introduced by the instrument maker, William Gaertner, from whom any of the special pieces used in the course may be obtained. Furthermore, in nearly all of the exercises, the quantity sought is obtained by two distinct methods and the results compared.

Finally, since the book represents a college, not a high school, course, the aim has been, not so much to acquaint the student with interesting and striking phenomena, as to give him an insight into the real significance of physical things—to introduce him to the very heart of the subject by putting him in touch with the methods and instruments of modern physical investigation, and by carrying him through the processes of close reasoning by which the present science of Physics has been developed. Students who enter upon the course are expected both to have completed a year of secondary school Physics, and to have gained some familiarity with the principles of Trigonometry.

The author's justification for the publication of this, the first book in the series, is the hope that the presentation of a method of instruction which has been found most satisfactory in the University of Chicago may not be altogether without usefulness, or at least suggestiveness, to teachers of Physics in other institutions. The custom usually followed is to hold lectures and quizzes upon a group of eight experiments, before taking up the laboratory work. The problems are of course invaluable aids in the fixing of principles. In the laboratory not more than twenty students are ever permitted in a single section. No more than two duplicate pieces of apparatus are ever used. Duplicates of the forms of record which have been inserted in the manual are filled out by the student and handed to the instructor as each experiment is completed. In addition to filling out these record slips each student keeps a systematic note-book, in which are entered in the laboratory, not at home, all observations and all calculations of whatever kind. This note-book is a complete record of the student's work, and should of course be so arranged as to be easily intelligible to anyone, even though he be unfamiliar with the course.

This book is the successor of *A College Course of Laboratory Experiments in General Physics*, published by the present director of the National Bureau of Standards, Professor S. W. Stratton, and the author. It is to Professor Stratton that the design of much of the apparatus is due. In the preparation of the present course the author has had the invaluable assistance not only of Professor Stratton, but also of Mr. G. M. Hobbs and Dr. H. G. Gale, both instructors in Physics at the University of Chicago.

UNIVERSITY OF CHICAGO,
August 27, 1902.

CONTENTS

MECHANICS

		PAGE
I.	Uniformly Accelerated Motion	9
II.	Force Proportional to Rate of Change of Momentum ($f = ma$)	15
III.	Composition and Resolution of Forces	21
IV.	The Principle of Work	29
V.	Energy and Efficiency	42
VI.	The Laws of Impact	52
VII.	Elastic Impact. Coefficient of Restitution	58
VIII.	Elasticity. Hooke's Law: Young's Modulus	65
IX.	The Coefficient of Rigidity	71
X.	Moment of Inertia	78
XI.	Simple Harmonic Motion	87
XII.	Determination of "g"	95
XIII.	The Law of Centripetal Force	100

MOLECULAR PHYSICS AND HEAT

XIV.	Boyle's Law	105
XV.	Density of Air	114
XVI.	The Measurement of Temperature	122
XVII.	Law of Avogadro—Densities of Gases and Vapors	138
XVIII.	The Pressure-Temperature Curve of a Saturated Vapor	152
XIX.	Hygrometry	164
XX.	Archimedes' Principle	172
XXI.	Capillarity	181
XXII.	Calorimetry	198
XXIII.	Expansion	215
Appendix		223
Index		

MECHANICS

I

UNIFORMLY ACCELERATED MOTION

Theory

Conceive of a body moving in a straight line with continually changing velocity. Its motion is said to be uniformly accelerated when it makes equal gains of velocity in equal intervals of time. The rate of change of velocity, or, for the present case, the gain in velocity per unit of time is called the acceleration of the body.

Uniform acceleration defined.

LAWS OF UNIFORMLY ACCELERATED MOTION.—The following laws are derived at once from the above definition:

1. If v represent the velocity of the body at the end of t units of time, a its acceleration and v_0 its velocity at the beginning of the t units, then

Velocity in terms of acceleration and time.

$$v = at + v_0; \qquad (1)$$

or, if the body start from rest (i.e. if $v_0 = 0$)

$$v = at. \qquad (2)$$

This law is nothing more than the mathematical statement of the definition.

2. If s_1, s_2, s_3, s_n, and s_{n+1} represent the distances traversed by the body during the 1st, 2d, 3d, nth and $(n+1)$th units of time respectively, then

Acceleration in terms of successive space intervals.

$$a = s_2 - s_1 = s_3 - s_2 = \ldots\ldots\ldots s_{n+1} - s_n \qquad (3)$$

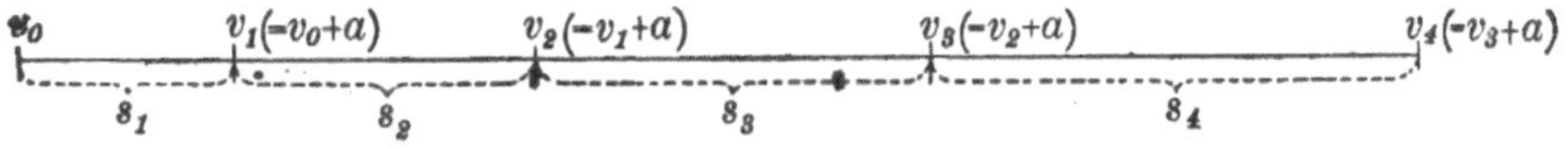

FIGURE 1

Proof.—Let the straight line (Fig. 1) be the path of a body moving with uniform acceleration, and let v_1, v_2, etc., be the velocities at the ends of the units of time 1, 2, etc., and s_1, s_2, etc., the spaces traversed during these units. The space passed over during any interval of time must always be the *mean velocity* multiplied by the number of units of time in the interval. In case the velocity increases uniformly this mean velocity is evidently the half sum of the velocities at the beginning and at the end of the interval. Hence, e.g. (See Fig. 1),

$$s_4 = \left(\frac{v_3 + v_4}{2}\right) \times 1 = \frac{v_3 + (v_3 + a)}{2} = v_3 + \frac{a}{2}. \tag{4}$$

Similarly $s_3 = v_2 + \frac{a}{2}$. $\therefore s_4 - s_3 = v_3 - v_2$. But $v_3 - v_2$ is by definition a. $\therefore s_4 - s_3 = a$. Similarly for $s_3 - s_2$, etc. Q. E. D.

The acceleration can therefore be most directly determined by measuring *distances* traversed in successive units of time.

Space in terms of acceleration and time.

3. If S represent the total space traversed during t units of time, then

$$S = \tfrac{1}{2} at^2 + v_0 t; \tag{5}$$

or, if the body start from rest

$$S = \tfrac{1}{2}at^2. \tag{6}$$

Proof.—Total space = mean velocity × time = $\frac{1}{2}$ (initial velocity + final velocity) × time = $\frac{v_0 + (v_0 + at)}{2} \times t = \frac{1}{2} at^2 + v_0 t$. Q. E. D.

Velocity in terms of space and acceleration.

4. If v represent the velocity of a body after it has moved over a space S with an acceleration a, and if v_0 represent the velocity which the body had at the point from which S is measured, then

$$v = \sqrt{2aS + v_0^2}. \tag{7}$$

or, if the body start from rest, (i.e., if $v_0 = 0$)

$$v = \sqrt{2aS}. \tag{8}$$

Proof.—From (1), $v = at + v_0 \therefore t = \frac{v - v_0}{a}$. Now if $\bar{v}$ be used to represent the average velocity with which the body traverses the space S, then $\bar{v} = \frac{v + v_0}{2}$, but $S = \bar{v}t = \frac{v + v_0}{2} \times \frac{v - v_0}{a} = \frac{v^2 - v_0^2}{2a}$. Hence solving for v,

$$v = \sqrt{2aS + v_0^2}; \text{ or if } v_0 = 0,\ v = \sqrt{2aS}. \qquad \text{Q. E. D.}$$

Experiment

bject. The object of this experiment is to study the motion of a reely falling body; in particular, (1) to ascertain whether a freely falling body moves with uniformly accelerated motion, and (2) to determine the acceleration of this motion centimeters per second.

The acceleration machine. The falling body is a brass frame *a*, (Fig. 2), which weighs about 1 kgm. and falls a distance of some 120 cm. through guides which offer very little friction. The cord, which is shown in the figure passing over the pulley and supporting balancing weights, is in this experiment detached from the frame *a*, so that the latter falls freely. The falling frame carries with it a tuning fork *b*, one prong of which is provided with a light stylus which, during the descent, traces a wavy line upon the blackened glass plate *M* (see Fig. 3). The frame is first raised to the top of the ways where it is held in place by an eccentric catch, which keeps the prongs slightly spread. Turning the lever *l* draws the eccentric up into the crossbar *o*, releases the prongs, and allows the frame and vibrating fork to fall. The plate *M* can be shifted sidewise so that a number of traces can be obtained. Two dash-pots at the base of the instrument catch the falling frame and take up the jar.

FIGURE 2

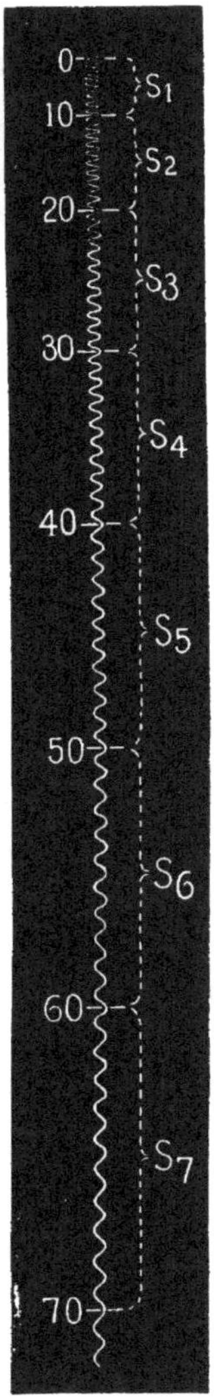

FIGURE 3

DIRECTIONS.—Hang a plumb-bob from *r* and adjust the leveling screws in the base until a strictly vertical fall is assured. Remove the glass plate M and smoke it evenly from top to bottom in a gum-camphor flame.

Production and measurement of the trace.

Replace it and adjust the stylus so that it will follow smoothly behind the falling fork, pressing very lightly against the plate. Then release the fork. Having obtained at least two good traces, remove the plate, set it in the horizontal wooden frame provided (see Fig. 4), and count off the waves by tens from top to bottom of each trace, marking

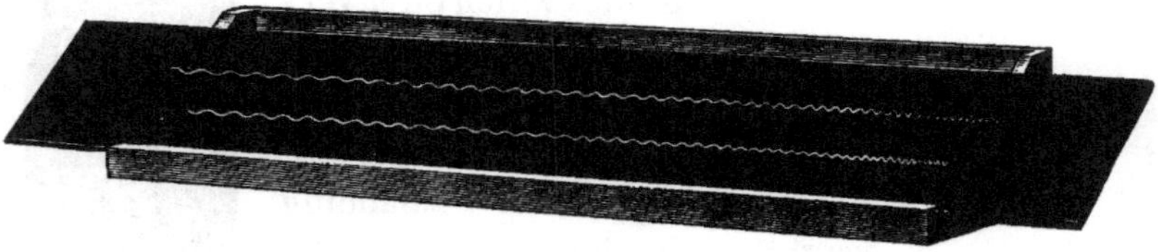

FIGURE 4

as accurately as possible the crest of each tenth wave by means of a pin-scratch (see Fig. 3). Next turn the plate over, smoked side down, and hold a meter stick on edge upon the unsmoked side of the plate just over the trace. It will then be easy to read upon the scale the distances between successive pin-marks.

Although the scale and the trace are the thickness of the glass apart, the error of parallax, i. e., the error in reading arising because the line of sight is not perpendicular to the plate, may be avoided entirely. For, since the blackened plate acts as a mirror, it is only necessary, in taking a reading, to place the eye so that the images of the scale divisions and the scale divisions themselves appear in the same straight line. In taking the readings estimate in each case to tenths mm.

Elimination of parallax.

By subtracting s_1 from s_2, s_2 from s_3, s_3 from s_4, etc. (see Fig. 3), obtain from each trace as many values of the acceleration as is possible, when the unit of time chosen is the time of 10 vibrations of the fork. Repeat the same operation when the unit is the time of 20 vibrations; then when it is the time of 30 vibrations; then of 40. If the motion is uniformly accelerated the values of a will not change as you move from the top to the bottom of the plate.

Acceleration for different units of time.

Law connecting "a's" and "t's". Value of g.

By comparison of the mean values of the accelerations for different units of time find the law which connects the numerical value of the acceleration with the length of the unit of time. From this law and the rate of the fork, as furnished by the instructor, determine g, the acceleration per second of the falling body.

Frictional resistance.

The true value of g is 980 cm. On account of the slight friction in the guides in addition to the air resistance, the value here found will fall a trifle short. From the total length of the fall and the correct value of g calculate what should be the total time of descent of the falling body. Compare this with the observed time obtained from the rate of the fork and the total number of waves upon the plate. The difference represents the retardation due to the friction of the air and of the guides. The difference between 980 and the observed value of g represents of course the negative acceleration due to friction. Since, when force is measured in dynes, force = mass × acceleration (see Ex. II), it is possible from an observation of the mass of the frame and fork to express the value of the retarding frictional force in dynes.

Record

	FIRST TRACE				SECOND TRACE			
	10 vib's.	20 vib's.	30 vib's.	40 vib's.	10 vib's.	20 vib's.	30 vib's.	40 vib's.
$s_2 - s_1$	——	——	——	——	——	——	——	——
$s_3 - s_2$	——	——	——	——	——	——	——	——
$s_4 - s_3$	——	——	——		——	——	——	
$s_5 - s_4$	——	——			——	——		
$s_6 - s_5$	——				——			
$s_7 - s_6$	——				——			
$s_8 - s_7$	——				——			
Mean	——	——	——	——	——	——	——	——
Means reduced *	——	——	——	——	——	——	——	——
Mean of means				——	Mean of means			——
Rate of fork = —— vib. per sec. ∴ g = ——							∴ g =	——

% of difference between the two values of g (observational error) = ——

% of departure of final mean from correct value of g = . . . ——

Total time of fall, calc'd value = —— obs'd value = —— % dif. = ——

Weight of frame and fork = —— ∴ No. dynes friction = ——

* Divide mean of 20's by 4, of 30's by 9, of 40's by 16.

Problems

1. Show that the law discovered experimentally above, viz., that the numerical value of the acceleration varies directly as the square of the unit of time, can also be obtained theoretically from the definition of acceleration, viz., the velocity (measured in cm. per second) which is gained during one second.

2. If the earth's period of rotation were to be suddenly doubled, and if a second of time were still to be defined as the same fraction of that period as at present, what effect would be observed in the numerical value of g?

Disregard all centrifugal tendencies.

3. If the earth's period were slowly changing, how could the change be detected?

4. Show that a body shot from the earth upward has at any given height the same velocity in the ascent as in the descent.

5. The Eiffel Tower is 335 meters high. How many seconds will elapse before an arrow, shot upward from the tower with a velocity of 4000 cm. per sec. reaches the earth?

II

FORCE PROPORTIONAL TO RATE OF CHANGE OF MOMENTUM ($f = ma$)

Theory

Newton's Laws of Motion.—The whole of Mechanics is based upon three great experimental principles first clearly stated in Newton's laws of motion. As will presently appear, the First and Second Laws contain but one experimental principle, the Third with the subjoined scholium contains two. Newton's statement of these laws is as follows:

Statement of laws.

I. Every body continues in its state of rest or of uniform motion in a straight line except in so far as it is compelled by force to change that state.

II. (Rate of) change of (quantity of) motion is proportional to force and takes place in the straight line in which the force acts.

III. To every action there is always an equal and contrary reaction; or the mutual actions of any two bodies are always equal and oppositely directed.

The First Law is completely contained in the Galilean definition of force, viz.: "*Force is that which changes the state of rest or uniform motion of a body.*"* It is therefore essentially a definition of force, and hence requires no proof.

Meaning of First Law.

The Second Law asserts that force is proportional to, and therefore may be measured by, *the rate of change of quantity of motion.* But "*quantity of motion,*" commonly called *momentum,* is the product of two factors, mass and velocity. *Rate of change of momentum* is therefore mass × rate of

Meaning of Second Law.

* Since the existence of forces in equilibrium is denied by this definition of force it is often found thus modified. "Force is that which changes *or tends* to change," etc. Newton, however, (and in this most modern writers follow him) regarded equilibrium as the balancing of *motions,* not of *forces.* Thus according to the Newtonian view each force actually produces its proper motion which is neutralized by the equal and opposite motion produced by the opposing force.

change of velocity, or mass × acceleration. The mathematical statement of the Second Law is therefore $f \propto ma$. This law is a definition of the measure of a force; but a little consideration will show that it is a definition which presupposes *experimental* knowledge, and consequently finds its justification in *experiment* only.

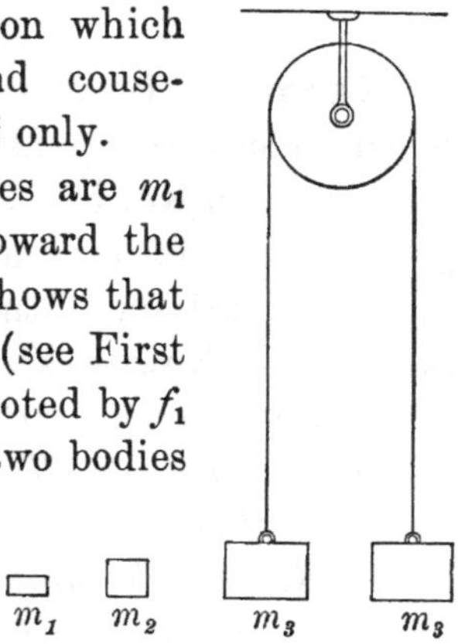

FIGURE 5

Proof of Second Law.

Thus, suppose two bodies whose masses are m_1 and m_2 (see Fig. 5) are dropped together toward the earth. The fact that they fall shows that certain forces act upon them (see First Law). Let these forces be denoted by f_1 and f_2. Now observation shows that the two bodies fall at the same rate, i. e., that they have a common acceleration. Call this acceleration g. Then the assertion made in the statement of the Second Law is that

$$\frac{f_1}{f_2} = \frac{m_1 g}{m_2 g} = \frac{m_1}{m_2}. \tag{9}$$

Next suppose that the two bodies are brought back to their original positions, where, necessarily, the same two original forces, f_1 and f_2, act upon them. Let now each body descend again, but this time let each be obliged to drag along with it some third mass which would otherwise remain at rest. Thus let m_1 be placed upon either one of the two equal masses m_3, which hang over a pulley of negligible weight and friction (see Fig. 5). In this case the moving force is evidently f_1, the mass moved is $m_1 + 2m_3$, and the acceleration is some measurable quantity a_1. Then let the masses m_3 be brought back to their original positions and m_2 placed upon one of them. The moving force is now f_2, the mass moved $m_2 + 2m_3$, and the acceleration some measurable quantity, a_2. The assertion made by the Second Law in regard to these two motions is

$$\frac{f_1}{f_2} = \frac{(m_1 + 2m_3)\ a_1}{(m_2 + 2m_3)\ a_2}. \tag{10}$$

Now (9) and (10) cannot both be true unless

$$\frac{m_1}{m_2} = \frac{(m_1 + 2m_3)\ a_1}{(m_2 + 2m_3)\ a_2}. \tag{11}$$

Since there is no sort of *a priori* ground for asserting the correctness of this relation, it is only the experimental proof of it which furnishes the justification for the Second Law. This proof can be made in any laboratory. But the most convincing evidence of the absolute correctness of the Second and Third Laws is furnished by astronomical observations. Predictions of eclipses made years in advance are wholly based upon these laws taken in connection with the law of gravitation.

UNITS OF FORCE.—If we arbitrarily choose as the absolute unit of force *the force which is required to impart to unit mass one unit of acceleration* the expression $f \propto ma$ changes to $f = ma$; for if, by definition, $f = 1$ when $m = 1$ and $a = 1$, then evidently $f = 4$ when $m = 2$ and $a = 2$, etc. The equation $f = ma$ is then merely the statement of Newton's Second Law. Note, however, that it holds in this form only when the force is expressed in the units defined above, i. e., in absolute units. If the unit of acceleration be taken as 1 cm. per sec., and the unit of mass as the mass of 1 cc. of water at 4° C. (i. e., 1 gram) the absolute unit of force is called a "*dyne.*" A dyne of force is then defined as *the force required to impart to 1 gram of mass an acceleration of 1 centimeter per second.*

Definition of the dyne.

There is another unit of force in common use, which is called the *gram of force.* It must not be confused with the *gram of mass.* The gram of force is defined as the force of the earth's attraction upon a gram of mass. Since this force is able to impart to the gram of mass an acceleration of 980 cm. per second, it is evident that one gram of force is equivalent to 980 dynes of force.

Number of dynes in a gram.

Experiment

Object. To prove that $f \propto ma$.

The apparatus is the same as that used in Ex. I, save that the cord c (Fig. 2) is attached to the frame and fork, carried over the pulley p and fastened at the other end to the weights m, one of which is adjustable. If the weights m just balance the frame and fork, the removal of any weight m_1 from m will cause the frame and fork to descend with an acceleration which may be measured as in Ex. I. The moving force will be m_1 grams, the mass moved will be the mass of frame and fork plus $(m - m_1)$.

Method.

After leveling the instrument as in Ex. I, adjust the mass m unti the fork, when given a slight downward movement, will just con tinue to move, but without acceleration. This elimi nates the friction of the pulley and guides. Nov remove a mass m_1 (say 400 gm.) from m and obtain two trace precisely as in Ex. I. Repeat when m has been diminished by second 400 gm. so that the moving force is now, say, 800 gm ($= m_2$). Measure the accelerations in the two cases according t the method given in Ex. I, using, however, but one unit of time say that of 20 vibrations of the fork. The mass of the frame an fork* as well as the masses m, m_1, and m_2 are to be determine by weighing upon a pair of trip scales. The test here made of th Second Law consists in verifying equation (11).

Directions.

Record

Weight of frame and fork (W) = ——. Balancing weight (m) = ——

	m_1 = ——				m_2 = ——			
	1st trace		2d trace		1st trace		2d trace	
	spaces	a_1	spaces	a_1	spaces	a_2	spaces	a_2
s_1	——		——		——		——	
		——		——		——		
s_2	——		——		——		——	
		——		——		——		
s_3	——		——		——		——	
		——		——		——		
s_4	——		——		——		——	
		——		——		——		
s_5	——		——		——		——	
		——		——		——		
s_6	——		——		——		——	
Means		——		——				
Final means		a_1 =		——		a_2 =		

$$\frac{f_1}{f_2} = \frac{m_1}{m_2} = \text{——}\dagger \quad \frac{[W + (m - m_1)] \times a_1}{[W + (m - m_2)] \times a_2} = \text{——}^* \qquad \% \text{ of error} = \text{—}$$

*The mass which, at the circumference of the wheel, would hav a moment of inertia (see Ex. X) equal to that of the wheel should theo retically be included in this weight. This is usually a negligible quantity

† Do not simply *indicate* the division. This blank is for the ***result*** o the division.

Problems

1. Over a weightless and frictionless pulley (see Fig. 5) are suspended masses of 200 gm. each. A weight of 100 gm. is added to one side. Find the acceleration thus imparted to the weights.

Take g, the acceleration of free fall, as 980; the acting force is then 100×980 dynes.

2. To the ends of a rope passing over a weightless pulley are attached two bodies of unequal masses, one of 200 gm., the other of something more than 200 gm. The acceleration imparted to the masses is observed to be 245 cm. per second. Find (1) the exact mass of the larger body; (2) the tension in the string. Express the tension both in dynes and grams.

SUGGESTION (1).—If $x =$ the difference between the two masses, then the acting force is xg dynes and the mass moved is $400 + x$.

SUGGESTION (2).—Since all freely falling bodies have the acceleration g, the force of gravity acting upon any mass of m grams is mg dynes. If, because of some retarding force (e.g. tension in a string), the *downward* acceleration is not g but some smaller quantity a, it is at once clear from the statement of the Second Law that the value of the *upward* or retarding force is $m(g - a)$ dynes. Thus, if the body is at rest, the upward force is mg dynes. If it has an upward acceleration amounting to a units the upward force is $m(g + a)$ dynes.

3. A 10 gm. mass is moving with a velocity of 40 meters per second. A force of 2000 dynes opposes its motion. How soon will it be brought to rest?

Find a from Second Law, then t from Ex. I.

4. A 900 kgm. projectile struck an embankment with a velocity of 400 meters per second. It penetrated 4 meters. Find the resistance in kgms. which the embankment opposed to its motion.

Find a from Ex. I.

5. A 50-lb. mass hangs from a spring balance in an elevator. How much does it appear to weigh at the instant at which the elevator begins to descend with an acceleration of 400 cm. per second? (1 kilo = 2.2 lb.)

See problem 2, suggestion (2).

6. A man pushes steadily upon a car weighing 1000 kilos. After 5 seconds it is moving with a velocity of .5 meters per second. Find the man's force in kilos.

Find a from Ex. I.

7. A man who can jump three feet high on the earth could jump how high on the moon? The mass of the earth is 80 times that of the moon. The diameter of the earth is $3\frac{2}{3}$ times that of the moon.

SUGGESTION.—The mathematical statement of Newton's law of gravitation is $f \propto \frac{mM}{r^2}$, in which f is the force acting between any two bodies, m and M their respective masses and r the distance between their centers of gravity. By the application of this law first find the relative values of the acceleration of a body at the surfaces of the earth and moon respectively, then note that the initial velocity produced by the spring is always the same.

III

COMPOSITION AND RESOLUTION OF FORCES

Theory

COMPOSITION OF FORCES—RESULTANT.—From the statement that "change of motion is proportional to, and in the direction of the impressed force," it follows, by implication at least, that a given force always produces the *same change* of motion in the direction of its action, whether the body upon which it acts is at rest or in motion, whether it is acted upon by other forces at the same time or not.

Implication of Second Law.

Consider, for example, a very short interval of time, say one-millionth of a second, during which a force f_1, acting in the direction AB (see Fig. 6), is accelerating a body at A. At the end of the interval the body will have acquired a velocity by virtue of which, during the second which begins at the close of this interval, it will move uniformly and in a straight line (see First Law) to some point B. If the interval of time be taken sufficiently small the acceleration during the interval may always be considered constant; and if, in all cases considered, the interval chosen be of the same length, then even for a variable force the acceleration will be simply the velocity acquired in one of these intervals divided by the length of the interval. Since, then, the velocity acquired in an interval is always proportional to the acceleration, the line AB may be taken as representing the acceleration a_1 due to the force f_1.

Effect of a single force. Meaning of acceleration.

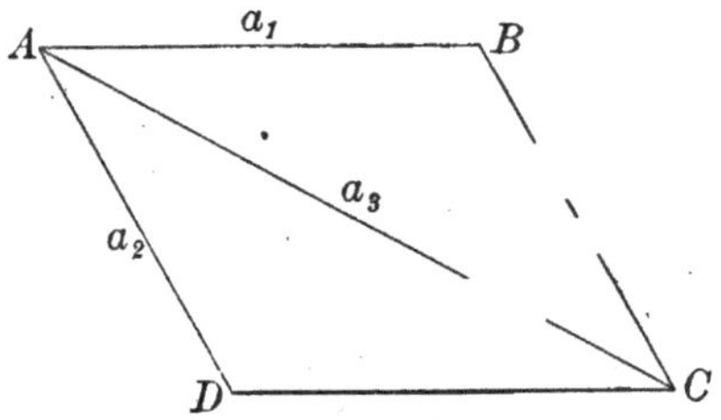

FIGURE 6

If, instead of receiving an acceleration represented in magnitude and direction by a_1, the body were acted upon by a force f_2 acting in the direction AD, it would receive, in the interval considered, a velocity which would carry it in one second to some point D. AD would then represent the acceleration due to f_2.

If, now, the two forces f_1 and f_2 act simultaneously, the Second Law asserts (by implication) that the effect of each force is the same as though the other did not act; i. e., that in the short interval considered, the body will receive a velocity which will carry it at one and the same time a distance AB in the direction of (i. e., parallel to) AB and a distance AD in the direction of AD. This is merely saying that the velocity acquired in the interval will carry the body in one second to C. Further, the path between A and C must be a straight line, because a body can move, by virtue of an acquired velocity, only in a straight line. Hence the line AC represents the joint, or resultant, acceleration due to the joint action of the two forces f_1 and f_2.

Joint effect of two forces.

But one single force f_3 can be found which, acting in the direction AC, would produce this acceleration a_3. Such a single force is called the *resultant* of the two given forces. Thus the *resultant* of any number of forces is defined as *that single force whose action would produce the same "change of motion" as is produced by the joint action of the several forces.*

Resultant force defined.

It is evident from the above, since, for a given mass, forces are proportional to accelerations, that the *resultant of any two forces acting upon a particle is represented in intensity and direction by the diagonal of the parallelogram the sides of which represent the two forces.* By the successive application of this rule to the case of the simultaneous action of *three or more* forces, such as a, b, c (see Fig. 7), the following general rule is obtained: *The resultant* (R) *of any number of forces is represented in magnitude and direction by the line which closes the polygon whose sides represent the several forces* when the latter are conceived as acting successively.

Resultant of any number of forces.

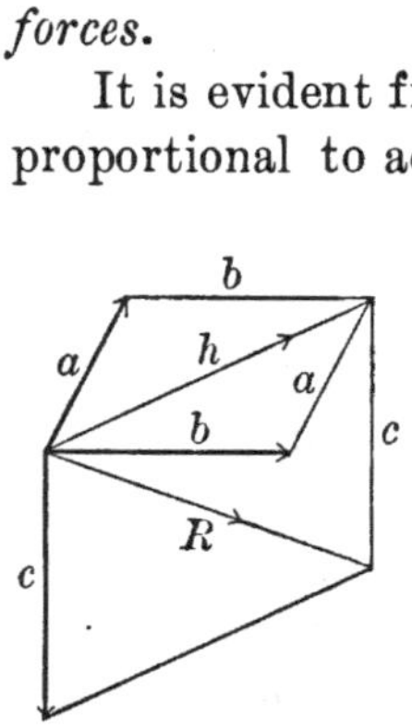

FIGURE 7

The problem, then, of finding the magnitude and direction of the resultant of two forces a and b which include an angle ϕ (see Fig. 8) resolves itself into the trigonometrical problem

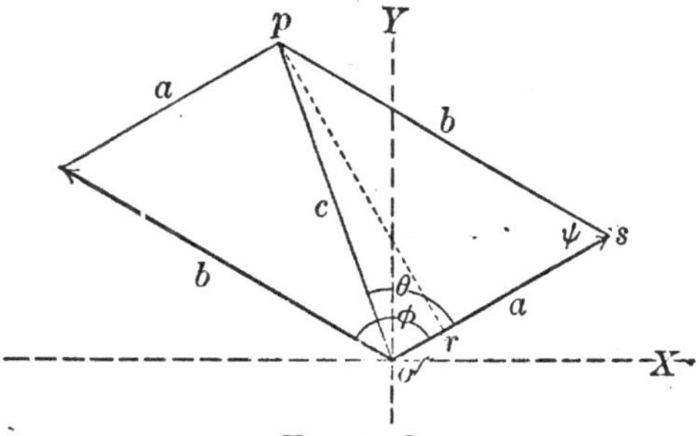

FIGURE 8

Calculation of resultant of two forces.

of finding the length of the line c and the value of the angle θ in terms of the given magnitudes a, b, and ϕ.

To find the length of c in terms of a, b, and ϕ, drop a perpendicular pr upon a.

Then $\qquad c^2 = \overline{or}^2 + \overline{pr}^2$. But $\overline{pr}^2 = b^2 - \overline{rs}^2$.

Hence $\qquad c^2 = \overline{or}^2 - \overline{rs}^2 + b^2$.

Now $\qquad \overline{or}^2 = (a - \overline{rs})^2 = a^2 - 2a\,\overline{rs} + \overline{rs}^2$.

Hence $\qquad c^2 = a^2 + b^2 - 2a\,\overline{rs}$.

But $\qquad \dfrac{rs}{b} = \cos\psi = -\cos\phi$ i. e. $\overline{rs} = -b\cos\phi$.

Hence
$$c^2 = a^2 + b^2 + 2ab\cos\phi. \qquad (12)$$

Again to find the angle θ in terms of a, b, and ϕ,

$$\frac{pr}{c} = \sin\theta \text{ and } \frac{pr}{b} = \sin\psi = \sin\phi.$$

Hence
$$\frac{b}{c} = \frac{\sin\theta}{\sin\phi} \text{ or } \sin\theta = \frac{b}{c}\sin\phi. \qquad (13)$$

The resultant of three or more forces may be found by combining the resultant of the first and second with the third, this resultant with the fourth, etc. A better method, however, is given in a following paragraph.

Definition of component.

RESOLUTION OF FORCES—COMPONENT.—*The component of a force in any specified direction is defined as that force which, acting in the direction specified, would produce the same effect, so far as motion in this direction is concerned, as is produced by the action of the given force.* Thus if AC (Fig. 9) represents the distance over which a body would move in one second by virtue of a velocity acquired through the action of the given force during one of the infinitesimal intervals above considered, AB evidently represents the distance moved over toward the east during this second; i. e., if AC represents the actual acceleration,

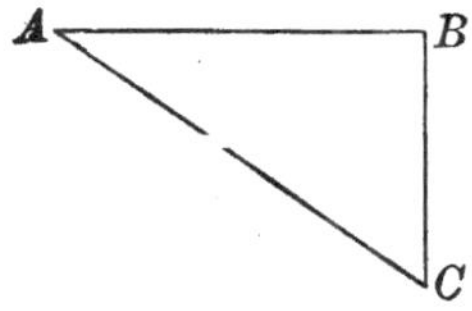

FIGURE 9

AB represents the *eastward* acceleration. Or, finally, a force represented by AB is equivalent, so far as motion toward the east is concerned, to the given force AC. AB is therefore the component of AC in the easterly direction.

Thus, if the body which is acted upon by the force AC is free to move only in the direction AB (conceive, for example, of a car on overhead rails pulled forward with the aid of a rope by a man on the ground), its motion under the action of the force AC must be precisely the same as though the force AB were acting instead of AC.

The process of finding the component of a given force in any direction consists, then, in finding the *projection* in the required direction of the line which represents the given force.

Calculation of component.

Thus AM, (Fig. 10) is the component in the direction AE of the force AF. But $AM = AF \cos \theta$. Hence, in general, the *component in any specified direction of a given force is the product of the given force into the cosine of the angle included between the specified direction and the direction of the given force.*

Just as AM is the component in the direction AE of the force AF (see Fig. 10) so AG is the component of AF in the direction AH. Now, if AE and AH are at *right* angles to each other, then the figure $AGFM$ is a parallelogram of which AF is the diagonal and AM and AG the sides. Hence, (see above under RESULTANT) the single force AF might entirely replace two forces represented by AM and AG. Conversely the single force AF must be in every respect *replaceable by* the two forces AM and AG. *This process of replacing a given force by its components in any two directions at right angles to each other is called the resolution of force.*

Definition of resolution of force.

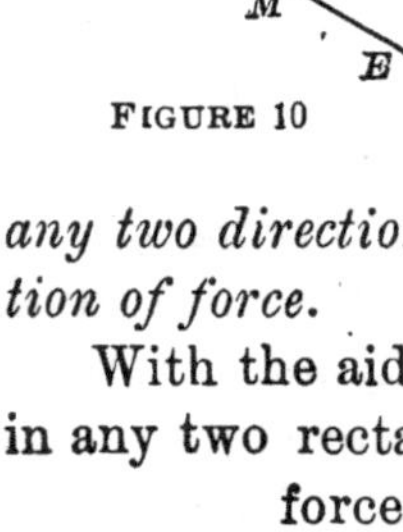

FIGURE 10

With the aid of the resolution of forces into their components in any two rectangular directions, the resultant of any number of forces may be easily obtained. Thus: Replace all of the forces F_1, F_2, F_3, etc. (see Fig. 11), by their components along any two rectangular axes X and Y. Find the algebraic sum $\mathbf{x}$, of all the x components x_1, x_2, x_3, etc. Similarly let Y represent the algebraic sum of all of

Calculation of the resultant of many forces.

the y components y_1, y_2, y_3, etc. The resultant of X and Y (see Fig. 11) will then be the resultant of the forces F_1, F_2, F_3, etc. The direction θ and the intensity R of this resultant may be obtained from any two of the following simple relations:

$$\left.\begin{array}{l} Y/X = \tan\theta \quad Y/R = \sin\theta \\ X^2 + Y^2 = R^2 \end{array}\right\} \qquad (14)$$

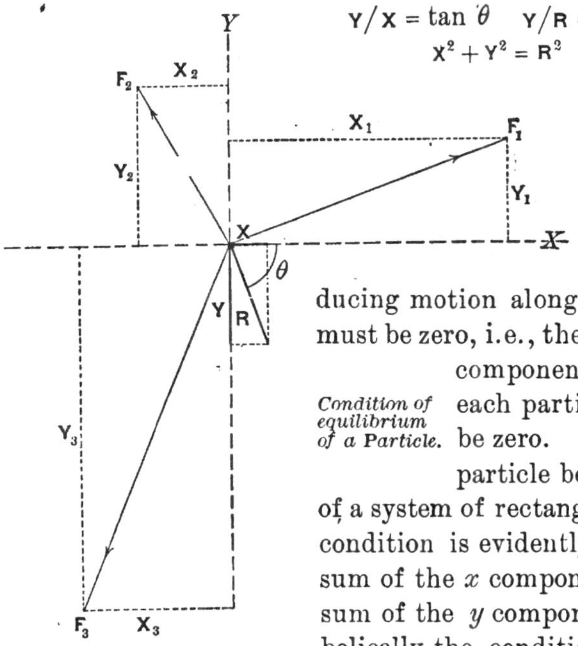

FIGURE 11

If a particle which is subjected to the action of forces is at rest, the total effectiveness of the forces in producing motion along any particular line must be zero, i.e., the algebraic sum of the components of the forces in each particular direction must be zero. If the position of the particle be taken as the origin of a system of rectangular coördinates this condition is evidently fulfilled when the sum of the x components is zero and the sum of the y components is zero. Symbolically the condition of equilibrium of a particle is therefore

Condition of equilibrium of a Particle.

$$\Sigma x = 0 \text{ and } \Sigma y = 0 \qquad (15)$$

in which Σ is the sign of summation.

Experiment

Object. To verify the laws of composition and resolution of force.

The apparatus consists of a horizontal force table (see Fig. 12) about the circumference of which may be set at any desired angles the four pulleys a, b, c, d, over which any desired weights may be hung. A pin holds the intersection of the cords in position at the center. When a test for equilibrium is to be made the pin is removed.

Method.

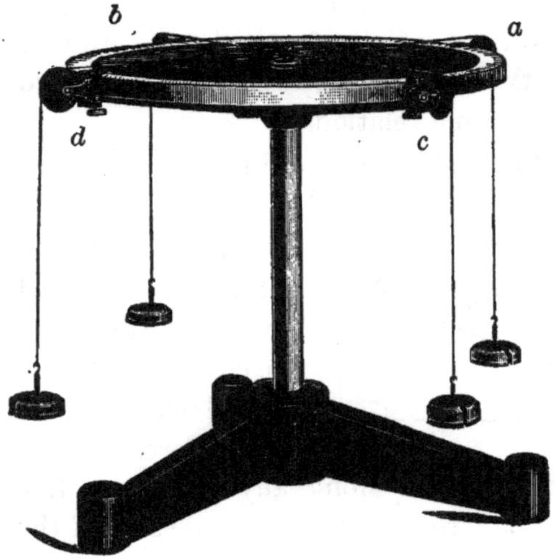

FIGURE 12

1. Set two of the pulleys at the angles marked 30 and 150, and apply weights of 100 gm. and 150 gm. respectively, remembering not to overlook the weights of the pans themselves. Then calculate, with the aid of equations 12 and 13, the direction and magnitude of the resultant. Set a third pulley 180° from the calculated angle, apply the calculated weight and remove the center-pin. To show that the equilibrium is not due to friction, displace the junction of the cords and tap the table. The junction should return to the center.

Resultant of two forces.

2. Resolve each of the three forces into its x and y components and verify the condition of equilibrium expressed in (15).

Condition of equilibrium.

3. Insert the center-pin, set three of the pulleys at any angles, and apply any weights. Calculate by means of *a resolution of all of the forces into their x and y components* the direction and magnitude of the force requisite to produce equilibrium. Test as in **1.**

Resultant of many forces.

4. Fasten one end of a spring balance to the hook in the wall at o (see Fig. 13) and the other end to the ring at the extremity of the stick mn. From the ring hang weights, and adjust the other end of the stick until it is perpendicular to the wall (see Fig. 13). Calculate the force in the spring balance and compare with the observed value. Do not overlook the weight of the stick itself.

Vertical component of oblique force.

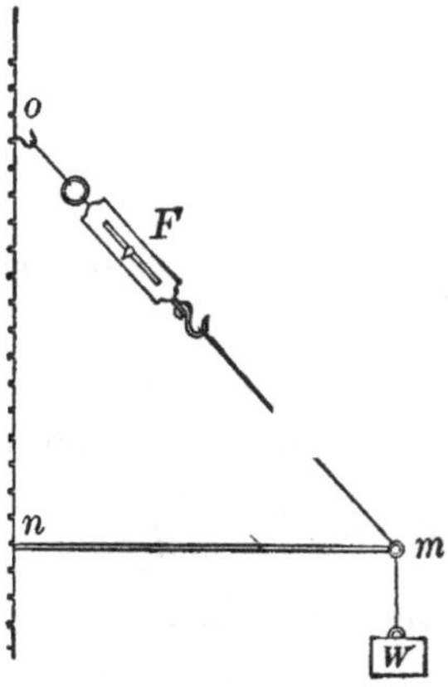

FIGURE 13

Record

1.

Forces	Direction	Magnitude
1st	———	———
2d	———	———
∴ Resultant	———	———

2.

Forces	x Comp.	y Comp.
1st	———	———
2d	———	———
3d	———	———
Sum	———	———

3.

Forces	x Comp.	y Comp.
1st	———	———
2d	———	———
3d	———	———
Sum	———	———

∴ Resultant magnitude = ———

∴ Resultant direction = ———

4. Force in spring bal. cal'd ———

Force in spring bal. obs'd ———

Problems

1. Show that the acceleration of a body sliding without friction down an inclined plane of length l and height h is $g \times \frac{h}{l}$.

2. Show that the velocity acquired in sliding down the plane is the same as that acquired in falling through the vertical height h, viz., $v = \sqrt{2gh}$.

3. By dividing the arc ab (Fig. 14) of a vertical circle into a large number of small inclined planes and applying the result of Problem **2**, show that a body sliding without friction down the arc ab acquires the same velocity as though it fell vertically from b to c.

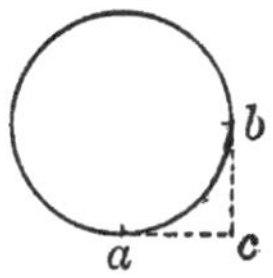

FIGURE 14

4. A 150-lb. man standing in the middle of a tight rope 60 ft. long depresses the middle 5 feet below the ends. Find the additional tension in the rope which is caused by his weight.

See Fig. 13 for suggestion as to method.

5. A bridge span is 5 meters high and 20 long (see Fig. 15). Find the vertical pressure and the horizontal thrust upon each of the piers per ton of weight at o.

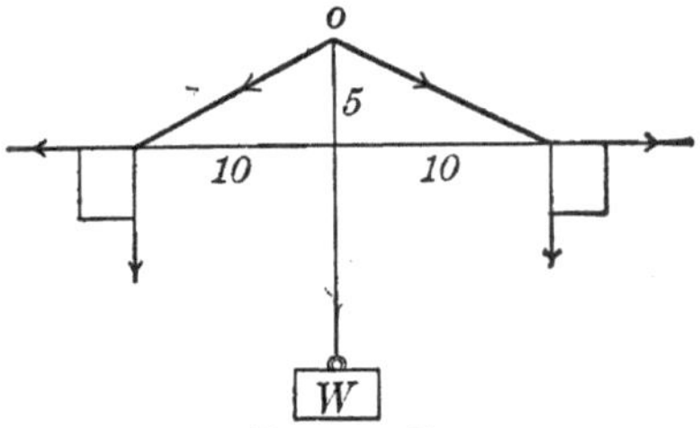

FIGURE 15

6. A bullet is fired from a gun with a velocity of 200 meters per second at an angle of 30° with the horizontal. How high will it rise? If the gun is on the surface of the earth at what distance from the gun would the bullet strike the earth if there were no air resistance?

7. If the gun mentioned in **6** were on the top of the Eiffel tower how far from the base would the bullet strike the earth? (Height of tower = 335 meters.)

8. Show that the time of descent down all chords which start from the top of a vertical circle is the same.

IV

THE PRINCIPLE OF WORK

THE LAW OF MOMENTS

Theory

SCHOLIUM TO THIRD LAW.*—The Third Law asserts that force is dual in its nature, that a change in the momentum of a body never takes place without an equal and opposite change in the momentum of some other body or bodies. The reasons for this assertion and some of its consequences are discussed in Ex. VI.

The assertion of the Third Law.

An additional and wholly distinct assertion is made in the scholium to the Third Law. It is that the work done by the force which produces the motion is always equal to the work done against the resistance to the motion. The scholium reads: "If the action (activity) of an agent be measured by its amount and its velocity conjointly, and if similarly the reaction (counter-activity) of the resistance be measured by the velocities of its several parts and their several amounts conjointly, whether these arise from friction, molecular force, weight, or acceleration, action and reaction in all combinations of machines will be equal and opposite." In other words, *the acting force F times the velocity v of its point of application equals the resisting force R times the velocity v' of its point of application. Symbolically* $Fv = Rv'$.

The assertion of the scholium to the Third Law.

Now the "*work*" W of a force is defined as *the product of the force acting F and the distance s through which its point of application moves during its action.* Thus the equation

The definition of work.

$$W = Fs \qquad (16)$$

is merely the definition of the quantity which is called work.

---

* See *Encyclopedia Britannica,* Article on "Mechanics."

Since velocity is merely rate at which space is traversed, i. e., space per second, it is seen that the assertion of the scholium is then simply that in all machines, during any given time, *the work of the acting forces is equal to the work of the resisting forces.* Symbolically,

The "work" principle.

$$Fs = Rs'. \tag{17}$$

This equation represents, then, the briefest possible statement of the "work principle," which is the essence of the scholium to the Third Law. The principle is a generalization from experiments upon all sorts of machines. It is, however, to be observed, that in all experiments in which there is acceleration the resistance to acceleration which the body offers because of its inertia must be included among the reacting forces, and must be put equal and opposite to that force whose single action would produce the observed acceleration, i. e., equal to ma, m being the mass of the body, and a its actual acceleration.

THE PRINCIPLE OF MOMENTS.—The great principle above asserted can of course be tested only for particular cases. One of the simplest of these is that of a rigid but weightless lever, free to rotate about a fulcrum. Fig. 16 represents the force diagram for the case of such a simple lever. F represents a force which is producing a clockwise rotation. W represents a force which is resisting this rotation.

Work principle applied to straight lever.

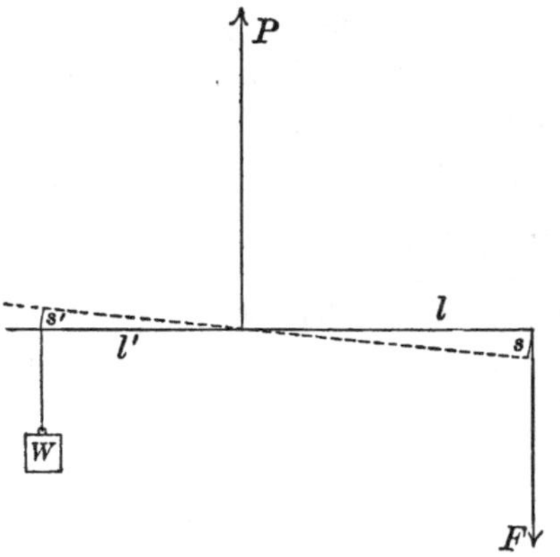

FIGURE 16

P is the upward force exerted by the fulcrum against the lever. If the fulcrum be a knife-edge, and the bar inflexible, there are no frictional or molecular forces. (Molecular force is such as is called into play by bending a bar, stretching a spring, etc.) If the motion be made so slowly that the acceleration may be considered zero, then the weight W is the only resistance, and the work principle gives $Fs = Ws'$ (see Fig. 16). The force P does not appear in this equation because its point of application does not move; hence the work of this force is zero. Now, from Geometry, $\frac{s}{s'} = \frac{l}{l'}$. Hence

$$Fl = Wl'. \qquad (18)$$

This equation is *rigorously* true only when the *acceleration is reduced to zero;* that is, *when the bar is at rest or is moving uniformly.* It represents, therefore, the *condition of equilibrium* of the bar.

Work principle applied to rotation of body acted upon by any number of forces in a plane.

Next consider any rigid body which is pivoted at o and acted upon by the forces F, W_1, W_2 and P (see Fig. 17). F is producing clockwise rotation, W_1 and W_2 are resisting this rotation, and P, the pressure of the pivot, has no influence upon the rotation, since its point of application does not move. Since the dircetion in which the force F acts is not parallel to the direction in which its point of application moves, the force which is effective in producing the motion along s, the direction in which this point of application of F must move, is not F but the component of F in the direction of s, viz., $F \cos \theta$. Similarly the effective resistances, i. e., the resistances in the directions s_1 and s_2 in which the points of application of W_1 and W_2 must move, are not W_1 and W_2, but $W_1 \cos \theta_1$, and $W_2 \cos \theta_2$. The work principle then asserts, provided there be no acceleration, that

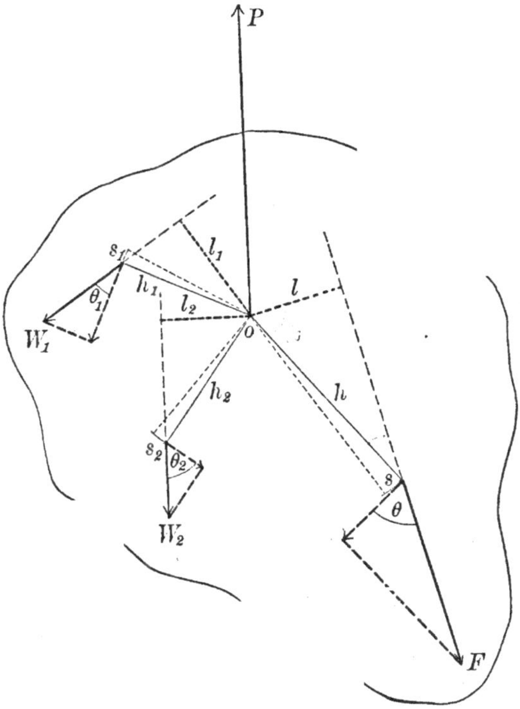

FIGURE 17

$$(F \cos \theta)\, s = (W_1 \cos \theta_1)\, s_1 + (W_2 \cos \theta_2)\, s_2. \qquad (19)$$

But from similar triangles, $s : s_1 : s_2 = h : h_1 : h_2$. Hence (19) becomes

$$(F \cos \theta)\, h = (W_1 \cos \theta_1)\, h_1 + (W_2 \cos \theta_2)\, h_2. \qquad (20)$$

But again, if perpendiculars l, l_1, l_2 be dropped from o upon the lines of direction of the forces (see Fig.), then $\cos \theta = \frac{l}{h}$, $\cos \theta_1 = \frac{l_1}{h_1}$, $\cos \theta_2 = \frac{l_2}{h_2}$. Hence (20) becomes

$$Fl = W_1 l_1 + W_2 l_2. \qquad (21)$$

This is therefore, according to the "work" principle (the scholium to the Third Law), the condition of no acceleration about o; i.e., *it is the condition of rest or uniform motion of rotation (equilibrium) of the body.*

Now the *lever arm* of a force is arbitrarily defined as *the perpendicular distance from the line of direction of the force to the axis of rotation.* Thus l, l_1 and l_2, are by definition the lever arms of the forces F, W_1 and W_2, respectively.

Definition of lever arm.

The "moment" of any force about any axis is defined as *the product of the force and its lever arm.* Thus Fl is by definition the moment of the force F about o.

Definition of moment.

Hence the condition of rotational equilibrium (21) expressed in words is "The *sum of the moments of the forces producing clockwise rotation must be equal to the sum of the moments producing counter-clockwise rotation;* or, if clockwise rotation be called negative, and counter-clockwise positive, the *algebraic sum of the moments of all the forces acting must be zero.* Symbolically

Condition of rotational equilibrium.

$$\Sigma Fl = 0. \qquad (22)$$

Although equation (22) has been developed only with reference to moments about the axis passing through o, nevertheless, if the pressure P be numbered among the acting forces the equation also holds for moments taken about any ideal axis conceived as passing parallel to the original axis through any point whatever in the plane of the forces. Thus if (see Fig. 18) the equation

Equation 22 holds for all axes.

$$Fl + W_1 l_1 + W_2 l_2 + Pl_o = 0 \qquad \text{(in which } l_o = 0\text{)}$$

be true for a body pivoted at o, then it is also true that, if o' be any point whatever in the plane,

$$Fl' + W_1 l'_1 + W_2 l'_2 + Pl_4 = 0.$$

This conclusion becomes self-evident from the consideration that, if the body be in equilibrium, P is the equilibrant of the three forces F, W_1, and W_2; i. e., so far as any effects produced by F, W_1 and W_2, are concerned, these three forces may be *replaced* by one *single* force equal and opposite to P applied at o. The moment, then, of this resultant force about any axis whatever must be equal and opposite to the moment of P; hence the sum of the moments of the forces which it replaces, viz., F, W_1, and W_2, must be equal and opposite to the moment of P.

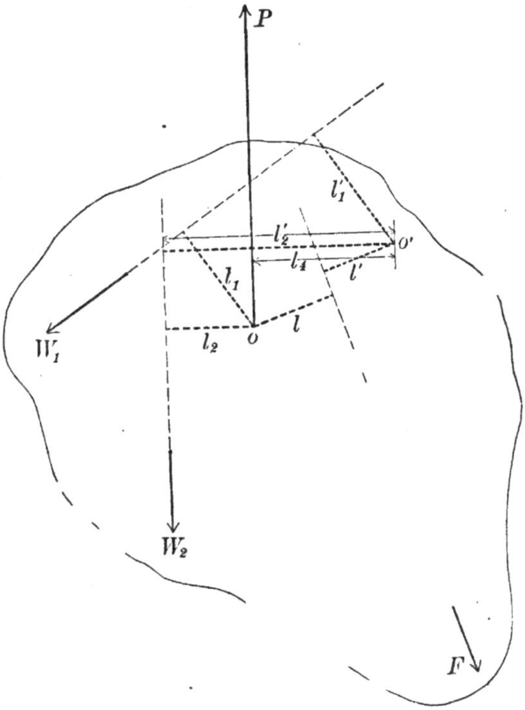

FIGURE 18

If a body of mass M be conceived as made up of equal infinitesimal particles each of mass m, then the center of mass of the body is defined as that point whose distance from each of three coördinate planes is the mean of the distances of all of the particles from each of these planes. Thus if n represent the total number of particles, and x_1, x_2 ...... x_n the distances of these particles from the YZ plane, then, from the above definition, the center of mass must lie in the plane whose distance x_c from the YZ plane is such that

Definition of center of inertia or center of mass.

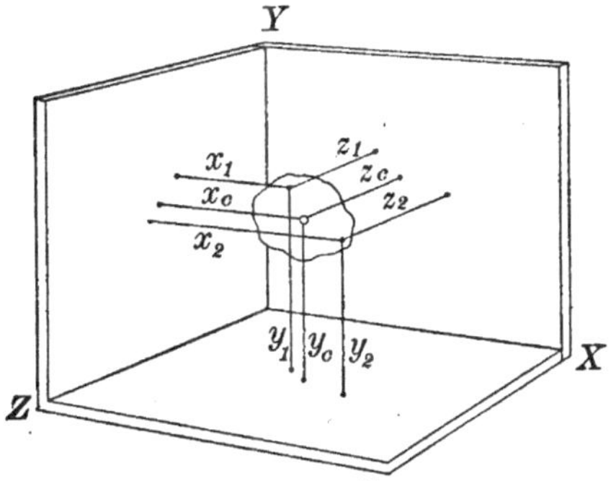

FIGURE 19

$$x_c = \frac{x_1 + x_2 + \ldots\ldots x_n}{n} = \frac{\Sigma x}{n} = \frac{\Sigma mx}{mn} = \frac{\Sigma mx}{M}. \qquad (23)$$

Similarly it must lie in planes whose respective distances y_c and z_c from the XZ and XY planes are such that

$$y_c = \frac{\Sigma my}{M} \quad \text{and } z_c = \frac{\Sigma mz}{M}.$$

The one point which can lie in all three of these planes is, by definition, the center of mass of the body. If the three coördinate planes be so chosen as to pass through this point, $x_c = 0$, $y_c = 0$, $z_c = 0$. Hence for this case

$$\Sigma mx = 0,\ \Sigma my = 0,\ \Sigma mz = 0. \qquad (24)$$

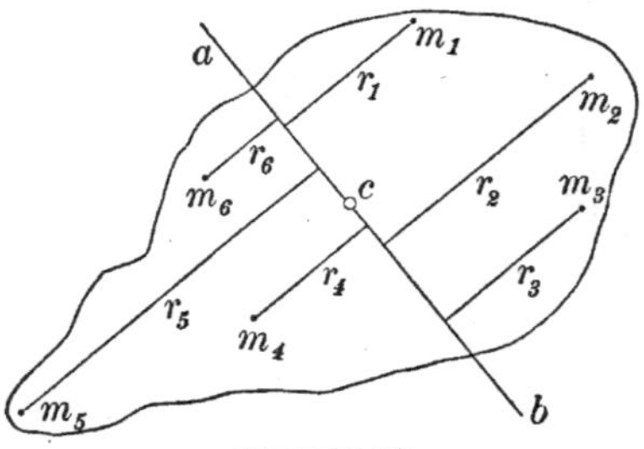

FIGURE 20

Or, if r be a general symbol representing the distance of a particle of the body from any plane whatever, the center of mass of the body is that point which is so situated that a plane may be passed through it in any direction whatever (see Fig. 20), and yet the following condition be always satisfied:

$$\Sigma mr = 0. \qquad (25)$$

Of course distances to one side of the plane are reckoned as positive, to the other side as negative.

If a body be placed in a uniform field of force,* equal forces will act upon all the equal elements of mass. Hence the point which satisfies the condition $\Sigma mr = 0$ must also satisfy the condition $\Sigma fr = 0$ (see Fig. 21).

Definition of center of gravity.

Now the *center of gravity* of a body is defined as that point which is so situated that, whatever the position of the body, a plane

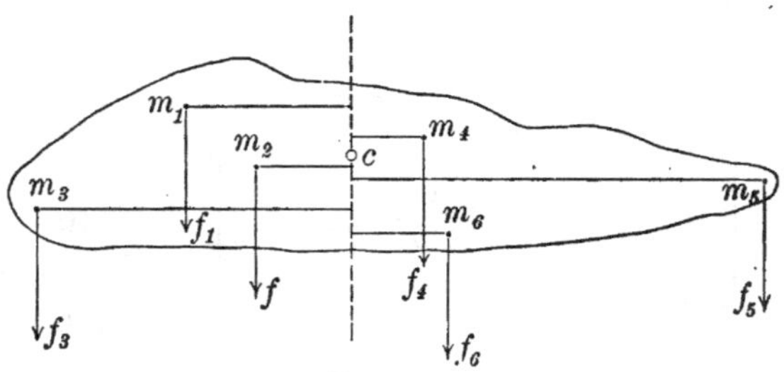

FIGURE 21

* A uniform field is defined as one in which the force acting upon a particle does not change as the particle moves about in the field.

which passes through this point and is parallel to the direction of the force always satisfies the condition

$$\Sigma fr = 0. \qquad (26)$$

It is evident that if a body has any center of gravity at all this point must always coincide with its center of mass; and yet a body possesses a center of gravity only when it lies in a *uniform* field of force. Its center of mass, however, is a perfectly definite point which depends in no way upon the location of the body.

Property of the center of inertia.

Now a rigid body placed in a uniform field must move without rotation, for every element tends to move at every instant with the same velocity. But the body would also move without rotation if a single force (of any intensity or direction) were applied to its center of inertia. For let ϕ_1, ϕ_2 etc. (see Fig. 22) represent the resistances to acceleration due to the inertias of the masses m_1, m_2, etc. Since for the center of inertia $\Sigma mr = 0$, and since inertias are proportional to masses, it follows that $\Sigma \phi r = 0$; i. e., the moments of the resistances due to inertia are always balanced about the center of inertia. Hence a single force applied at the center of inertia can never cause rotation. It is evident, then, that a single force applied at the center of inertia will cause exactly the same motion as is produced by the uniform field. This force must, of course, be equal in intensity to the sum of the forces exerted by the field upon all of the particles of the body. Hence, so far as all effects produced are concerned, it must be possible to *replace* the action of the field by the action of such a single force. It is therefore customary *to treat a body which is acted upon by gravity as though it were under the action of but one force, that force being applied at its center of gravity and being equal in intensity to its weight.* Thus the center of gravity is sometimes defined as that point at which the weight of the body may be conceived as concentrated.

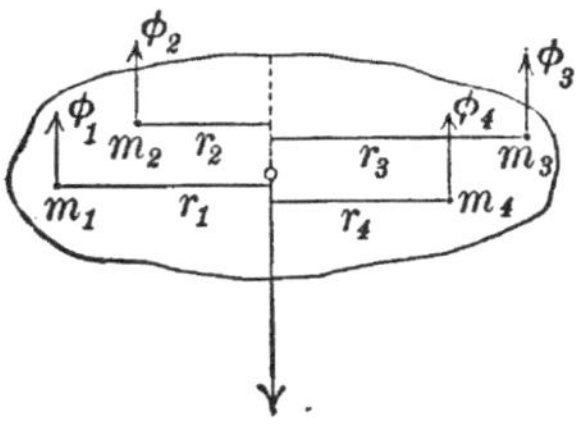

FIGURE 22

Experiment

Object.

To find the center of gravity of a body; to verify the principle of moments; and to study the theory of weighing.

The apparatus consists of a meter bar furnished with three knife-edges *a*, *b*, and *c* (see Fig. 23). *a* and *c* are attached to a metal frame through which the beam may be slipped and to which it may be clamped by means of the screw *s*. *c* is adjustable in a vertical plane; *b* is fixed to the bar. The beam is also provided with notches *m* and *m′*, and

The model balance.

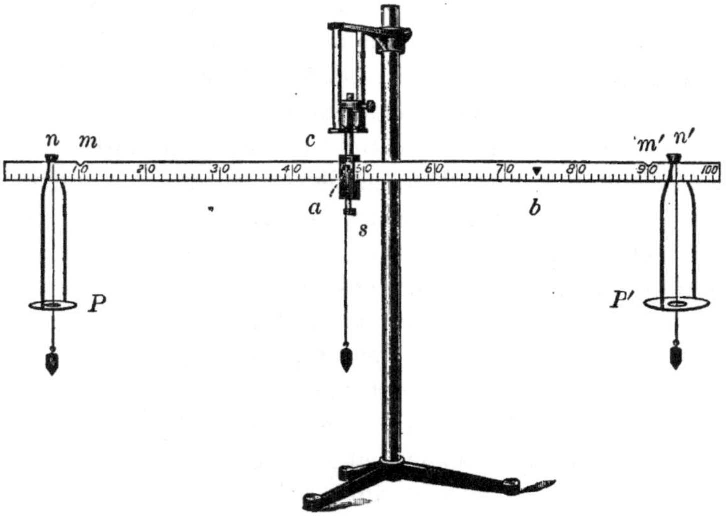

FIGURE 23

with pans *P* and *P′* which are supported from the top of the bar by knife-edges *n* and *n′*. A small hole passes vertically through each of the knife-edges *c*, *n*, and *n′*, so as to permit plumb lines to be dropped from these points.

DIRECTIONS.—**1.** Set the knife-edge *c* about 1 cm. above *a*. Support the beam from *c* and adjust the bar, by slipping it through the frame, until it balances in a horizontal position. Then hang *P* and *P′* from the bar, well out toward the ends, and slide one of them along until the bar is again horizontal. Read off the lever arms upon the graduated bar and write down the equation of moments. Now add 100 gm. to *P*, and, without altering the position of *P′*, slip *P* toward the fulcrum until a balance is again found. Write again the equation of moments. The solution of the two equations thus obtained will give the weights *P* and *P′*. Check the results by weighing on the trip scales.

To find weights of pans.

To find center of gravity of system.

2. Remove the pans, support the beam this time from a, and adjust the movable knife-edge and the bar until the whole system is in neutral equilibrium about a; i. e., until the system shows no tendency to move out of any position in which it is placed. This amounts to bringing the center of gravity of the system into coincidence with a. A small bit of soft wax, which may be attached to the bar at any desired point, greatly facilitates this adjustment.

To find weight of beam.

3. (a) Having brought the center of gravity of the system into coincidence with a, support the beam from c, hang the pans from the notches and place such weights upon them that the beam assumes a position inclined about 45° to the horizontal (Fig. 24). The equation of moments now involves the three forces W_1, W_2, and the weight of the beam W_3, and the three corresponding lever arms l_1, l_2, l_3. These latter are to be measured by bringing a half meter rule, which rests horizontally upon some firm support, up to the plane of the plumb lines which hang from the knife-edges. The equation of moments contains but one unknown quantity, viz., the weight of the beam. Solve for this unknown quantity.

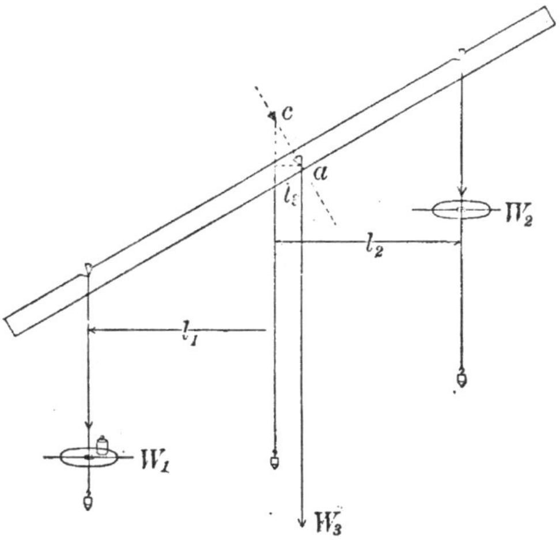

FIGURE 24

(b) Check the weight thus obtained by supporting the bar upon the knife-edge b and producing a balance by means of a known weight W* (see Fig. 25).

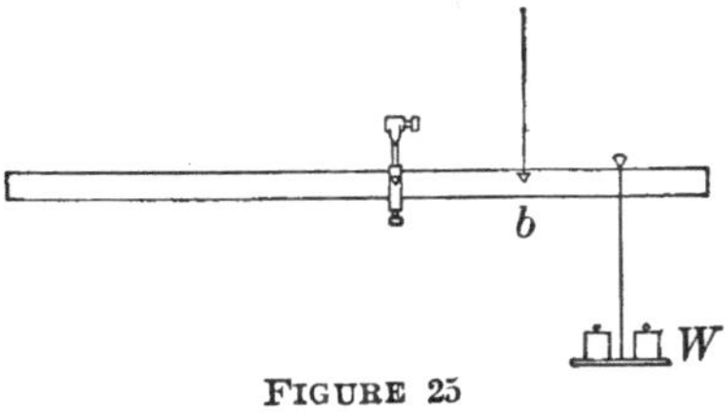

FIGURE 25

* The weights of the pans and plumb bobs should be included in W_1, W_2, and W_3 as they occur.

(c) As a final check weigh the beam upon the trip scales.

4. (a) *When the knife-edges n, n′, and c are not in the same straight line.* The "sensitiveness" of a balance is defined as the displacement produced when some arbitrarily chosen small weight is added to one pan. It should decrease as the load upon the pans increases, provided the supporting knife-edge c is above the line connecting the pan knife-edges n and n' (see Problems **2** and **3**, pp. 39, 40). Set the knife-edge c three or four centimeters above the line connecting n and n'. Support the system from c. Hang the pans from the beam, but not in the notches (a little friction at the knife-edges n and n' vitiates completely the results of this experiment), and bring the beam into the horizontal position. Set up a vertical meter stick behind one end of the beam and read off upon it the exact deflection produced by adding very carefully one gram to one of the pans. Then add 300 gm. to each pan. If the equilibrium is destroyed, reëstablish it by sliding along the pan to which the small weight was not added, and again take the sensitiveness. Repeat both observations to make sure that the results can be duplicated.

Sensitiveness for different loads.

(b) *When the knife-edges n, n′, and c are in the same straight line.* In this case the sensitiveness should be independent of the load (see Problem **2**, p. 39). Lower the knife edge c until it is in line with n and n'. Observe as in **4** (a). (The bending of the beam may still cause a slight dependence of sensitiveness upon the load.)

5. Hang unequal pans P and P' in the notches n and n', and slip the bar through the frame until a balance is obtained. The arms of the balance are now unequal, the pans are of unequal weight, and the center of gravity of the beam is not beneath the point of support. Nevertheless, if an unknown weight W be placed in the pan P and the

To make a correct weighing with false balances.

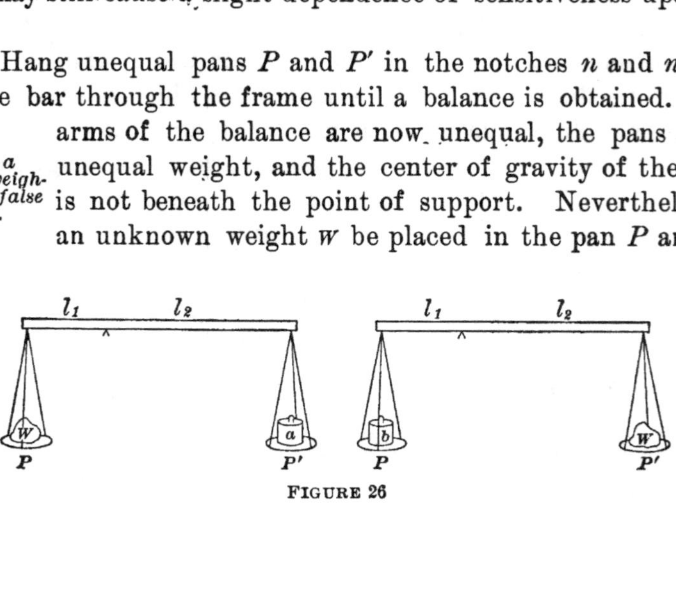

FIGURE 26

beam brought back to its original position by the addition of a known weight a to the other pan, the following equation must hold: $wl_1 = al_2$ (see Fig. 26). Now place w in P' and balance by means of a known weight b placed in P. The equation which now holds is $wl_2 = bl_1$. The solution of these two equations for w gives $w = \sqrt{ab}$. If a and b have nearly the same value, it is sufficiently correct to write $w \doteq \frac{a+b}{2}$. Hence all the errors of a balance are eliminated by a *double weighing*.

Record

1. Reading at a in condition of balance........................ = ——

Read'g for P = —— for P' = —— for $P + 100$ = —— ∴ P = —— P' = ——

By trip scales P = —— P' = —— % error in Ps = —— in P's = ——

3. (a) W_1 = —— W_2 = —— l_1 = —— l_2 = —— l_3 = —— ∴ W_3 = ——

(b) W = —— lever arm of W = —— of W_3 = —— ∴ W_3 = ——

(c) By trip scales W_3 = —— % error in (a) = —— in (b) = ——

4. No load, (a)	Reading before adding 1 gm.	——	after	——	dif. ——
" " (b)	" " " "	——	"	——	" ——
With load, (a)	" " " "	——	"	——	" ——
" " (b)	" " " "	——	"	——	" ——

5. a = —— b = —— ∴ w = ——. By scales w = —— % error = ——

Problems

1. Explain why a balance beam returns to a horizontal position when displaced therefrom.

2. Fig. 27 represents the case in which the three knife-edges n, n', and c are in the same straight line. Show from this figure that if in the horizontal position the center of gravity g of the beam be directly beneath c, i. e., if $Pl_1 = P'l_2$, then in the displaced position due to the addition of a small weight p to P, the new equation of equi-

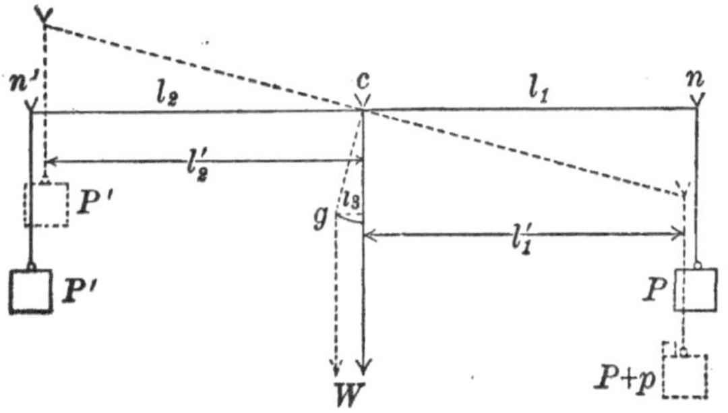

FIGURE 27

librium, viz.:, $(P+p)l'_1 = P'l'_2 + Wl_3$, reduces to $pl'_1 = Wl_3$. Hence show that the sensitiveness, which is the displacement of the beam produced by p (a quantity of which l_3 may be taken as the measure), is independent of the load, i. e., of P and P'.

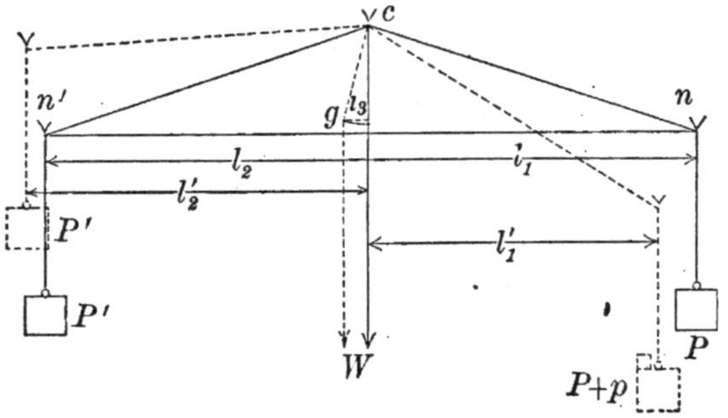

FIGURE 28

3. Show from Fig. 28 that when n, n', and c are not in the same straight line, then, after displacement, Pl'_1 is not equal to $P'l'_2$, and hence that the sensitiveness, i. e., l_3, depends upon the load, decreasing with the load if c be above the line nn', increasing with the load if c be below the line nn'.

4. Does the period of vibration of a balance vary with the load? If so, how and why?

5. With a given load how should the period be affected by a diminution in the distance between the knife-edge c and the center of gravity g?

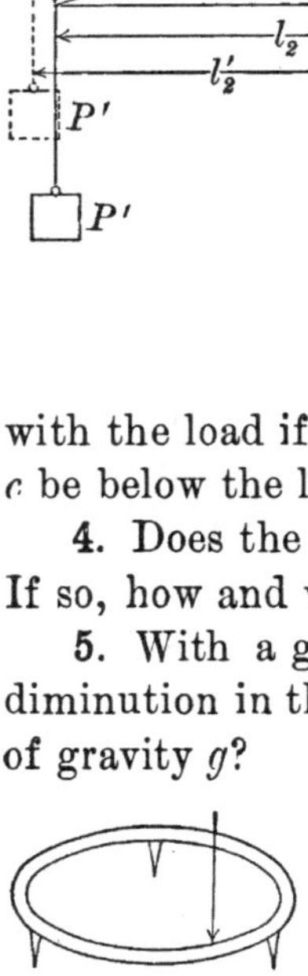
FIGURE 29

6. A circular ring weighing 5 lb. rests horizontally upon three points of support 120° apart. What is the least downward force, applied to the ring in a direction perpendicular to its plane, which will cause it to leave one of the points of support? (See Fig. 29.)

7. A uniform bar weighing 10 lb. is pivoted at one end (see Fig. 30). A horizontal force of 5 lb. is applied to the free end. When a condition of equilibrium is reached, what angle does the bar make with the horizontal?

FIGURE 30

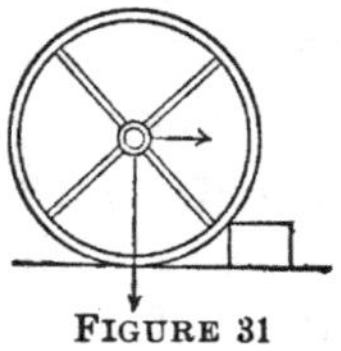
FIGURE 31

8. The axle of a wheel carries a load of 500 kilos. What horizontal force must be applied to the axle to raise the wheel over an obstacle 12 cm. high, the radius of the wheel being 50 cm.? (See Fig. 31.)

9. A uniform board 3 feet square and weighing 25 lb. rests on a block at *a* (see Fig. 32), and is kept from falling by a horizontal force at *B*. Find the force at *B* and the vertical and horizontal pressures upon the block at *a*.

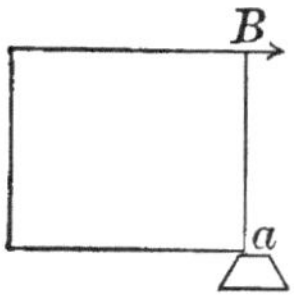

FIGURE 32

10. Two men carry between them a weight of 50 kilos. supported upon a uniform bar weighing 30 kilos. Where must the load be placed in order that one man may carry twice as much as the other?

Choose some convenient point as axis and apply the equation $\Sigma Fl = 0$.

11. A man wishes to overturn a cubical block which weighs 100 kilos and has a 5 ft. edge. In what direction and with what force must he push in order that he may accomplish his object most easily? After the block has once started will the required force increase or decrease? Why?

V

ENERGY AND EFFICIENCY

Theory

Definition of energy.

In the preceding experiment *work* was defined as the product of the force acting and the distance through which it moves the point to which it is applied. Symbolically, $W = Fs$.

The *energy* of an agent is defined as its *capacity for doing work.* Energy and work must of course be measured in the same units; yet it is obvious that they are not synonymous terms, for a body may possess energy and yet never apply it to the production of work. Work is done only when energy is *expended.*

Units of energy and work.

Since work is a product of force and space, the work unit must of course involve a force unit and a space unit. The absolute centimeter-gram-second (c. g. s.) unit of energy or of work is the dyne-centimeter, also called the *erg.* An erg of work is done when a dyne of force moves the point upon which it acts through a distance of one centimeter. Other work units are the gram-centimeter, kilogram-meter, foot-pound, etc., the definitions of which are evident from their names. Since a gram of force is 980 dynes, it is evident that a gram-centimeter is equal to 980 ergs.

Kinetic energy. Resistance to motion is inertia alone.

It was shown in Ex. IV that Newton's interpretation of the Third Law as given in the scholium is equivalent to the statement that in all mechanical operations the energy expended by the agent is equivalent to the work done against the four resistances, friction, molecular force,* gravity, and inertia. If the first three resistances are absent the only effect of the force is to impart *velocity* to the body. By virtue of this velocity the body itself becomes possessed of the capacity for doing work, for it can now move itself against a fric-

* Electrical and magnetic forces are here classed as molecular.

tional resistance, compress a spring, raise itself against gravity, or by impact overcome the inertia of some other body. *This energy which a body possesses because of the motion which has been communicated to it is called "kinetic energy."*

That the kinetic energy imparted to a body by the action of a force is exactly equal to the work done upon it, and that this is equal, in the absolute system of units, to one-half the mass of the body into the square of its velocity, may be shown as follows:

Proof that $KE = \frac{1}{2}mv^2$.

Let a body of mass m acquire a velocity v under the action of a constant force F acting for a time t, and in that time moving the body a distance s. The work done upon the body is then by definition Fs. Now let the body be brought to rest by being subjected to the action of an oppositely directed constant force F' which requires a time t' and a space s' in order to destroy the velocity v. The "kinetic energy" of the body, i. e., the work which it is capable of doing because of its velocity, is by definition $F's'$. But since every force is measured by the rate of change of momentum which it produces, the first force F is measured by the rate at which it imparts momentum, and the second force F' by the rate at which it destroys momentum, i. e.,

$$F = ma \text{ and } F' = ma'. \qquad (27)$$

Now, since F and F' are both constant forces, i. e., forces which produce uniformly accelerated motion, by Ex. I

$$s = \tfrac{1}{2}at^2 \text{ and } s' = \tfrac{1}{2}a't'^2. \qquad (28)$$

Therefore from (27) and (28)

$$Fs = \tfrac{1}{2}ma^2t^2 \text{ and } F's' = \tfrac{1}{2}ma'^2t'^2 \qquad (29)$$

But by hypothesis the velocity imparted by F and the velocity destroyed by F' are one and the same velocity. Hence

$$v = at \text{ and } v = a't'. \qquad (30)$$

It follows from (29) and (30) that

$$Fs = \tfrac{1}{2}mv^2 \text{ and } F's' = \tfrac{1}{2}mv^2.$$

Hence

$$Fs = F's' = \tfrac{1}{2}mv^2. \qquad (31)$$

Q. E. D.

If F and F' are *variable* forces it is only necessary to conceive them as made up each of the same number of very small elements, each element being a constant force. v will then be the velocity gained under the action of one of these constant elements of F and destroyed under the action of the corresponding element of F'. Hence the above demonstration is perfectly general.

Potential energy. Resistance is gravity or molecular force alone.

Next suppose that the resistance which the working force experiences is gravity alone, as when a body is taken from position a (see Fig. 33), and placed upon a hook in position b. The work done upon the body is the force of the pull times the distance ab. The pull is a variable force, being a little greater than the weight of the body when the motion is starting, and a little less when it is stopping. The kinetic energy which is imparted to the body during the first instants, when the velocity is being acquired, is all given back during the last instants when the velocity is being lost. When the operation is taken as a whole no velocity is imparted; hence the only resistance is gravity. Then by Newton's interpretation of the Third Law the work which has been done upon the body is equal to the work done against gravity, viz., the force of gravity upon the body times the distance ab. But, in this case, as in the preceding in which the resistance was inertia alone, the work may all be regained without the expenditure of any more work on the part of the agent. For, in returning to position a, the body is capable of lifting through the height ab any other body, e.g. body 2 (see Fig. 33), whose weight does not exceed its own. The ability to do work which the body possesses by virtue of its position at b is called its *potential energy*.

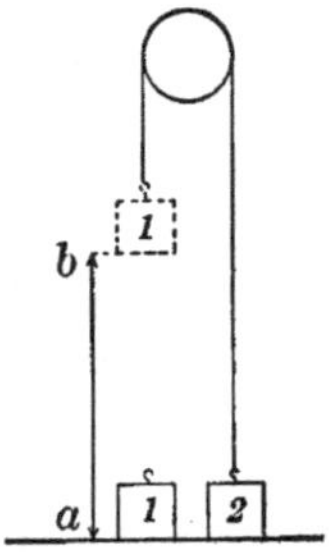

FIGURE 33

If the work be done against molecular force alone, as when a perfectly elastic spring is compressed, the work done can all be regained by releasing the spring, which, when compressed, is possessed of potential energy. *Potential energy is in general any energy which is put into a system by a change in the position of its parts.* Thus when the resistance is gravity or molecular force alone, an amount of potential energy equal to the work done is stored up; when the resistance is inertia alone, kinetic energy equal in amount to the work done appears.

But when the resistance is friction, the work done by the agent *can not* be regained. In Newton's day it was supposed to have disappeared altogether; but about the middle of the nineteenth century it was proved by Joule that for every erg of work which so disappears there always appears a perfectly definite quantity of heat; hence it is now customary to say that the work expended has been *transformed* into heat energy.

Heat energy. Resistance is friction alone.

The experiments of Joule, whereby the principle of the equivalence of heat and work was established, consisted in transforming heat into work in as large a variety of ways as possible, e. g., by means of the friction of different sorts of substances, by percussion, by compression, by the generation of electric currents the energy of which was finally dissipated in heat, etc. When the experiments were so arranged that the heat generated was taken up by a given quantity of water, it was observed that a given expenditure of mechanical energy always produced the same rise of temperature in the water. These experiments had much to do with securing general acceptance for the principle of *conservation of energy*, a principle blindly grasped at by philosophers from the earliest times; first stated, in the scholium to the Third Law, by Newton in 1687, but with respect to *mechanical* operations only; first asserted as a principle of universal applicability by the German physician J. R. Mayer in 1845; first generally accepted and universally recognized as the most fundamental and most fruitful principle in all physical science, after Joule, by his series of experiments extending from 1843 to 1878, had demonstrated the *equivalence of heat and work*. The present accepted value of the mechanical equivalent of heat, i. e., the number of ergs of work required to raise 1 gm. of water at 15° C. through 1° C. is

Origin of principle of conservation of energy.

$$4.19 \times 10^7.$$

But the principle of conservation of energy is more than the assertion of the equivalence of heat and work. It may be stated thus: *Every physical (or chemical) change of condition has a fixed mechanical equivalent*, i. e., *can be equated, under all circumstances, to one and the same amount of mechanical work.* In other words, whenever a change takes place in the condition of a body because of the expenditure upon it of mechanical energy (kinetic or potential), the change is equivalent to the work done, in the sense that if the body can be

Statement of principle of conservation of energy.

brought back to its original condition the whole of the energy expended may be *regained* either in the form of work or the equivalent heat.

Thus, applying the principle to a mechanical problem, it asserts at once that the kinetic energy of a moving body is equal to the work done in setting it in motion.

Illustrations of principle.

Applying it to a chemical problem it asserts, since the burning of 1 gram of carbon, i. e., the formation of carbon dioxide from carbon and oxygen, generates enough heat to raise 97,000 gm. of water through 1° C., that, if it were possible to directly pull apart the united carbon and oxygen atoms, $97,000 \times 4.19 \times 10^7$ ergs of work would be required to secure 1 gram of carbon from this compound. Applied to an electrical problem the principle asserts that if it requires 1000 kilogram-meters of work per second to drive a dynamo, then the work which this dynamo does per second in the motors which it runs, plus the mechanical equivalent of the heat developed in all of the machines, and in all the connecting wires must be exactly equal to 1000 kilogram-meters.

The principle is perhaps the most important generalization which has ever been made. It is merely an extension of the scholium to the Third Law, and, like it, rests upon universal experience rather than upon any one particular experiment. The only kind of direct test of which it is capable is of the kind which Joule made, and consists in mechanically producing a given change of condition (e. g., a given rise in the temperature of water) in as large a variety of ways as possible, and observing whether the work required always comes out the same. Of course, such an experiment tests the law only for one particular kind of physical change.

Basis for assertion of principle.

In all mechanical devices the work which the machine accomplishes is inevitably less than the work which is put into it, for the reason that there is always some friction and hence a part of the applied work disappears as heat. *Efficiency* is defined as the ratio of the work done *by* the machine in any given time to the energy expended *upon* it in the same time.

Efficiency.

Experiment

Object. To determine the efficiency curve (1) of a system of pulleys, (2) of a water motor.

FIGURE 34

Directions.

1. First weigh the movable block of pulleys and each of the pans (see Fig 34). Then give to W, which includes the weight of pan and movable block, successive values of about 300, 600, 1000, 1400, 1800 grams, and find the corresponding values which must be given to P, including pan, in order to produce *unaccelerated* downward motion of P. Calculate the efficiency corresponding to each case, and plot a curve with efficiencies as ordinates and loads as abscissae. Since the efficiency is the ratio of the *work* of the force P and the *work* of the force W, a determination of efficiency must involve a determination of the ratio of the distances through which the points of application of P and W move. This can be obtained without a measurement, as a little consideration will show, from the number of strands between which the weight of W is divided.

Energy expended upon motor.

2. In order to determine the energy expended upon a water motor in any time, it is necessary to know, first, the pressure p under which the water issues from the orifice o (see Fig. 35); second, the volume of water V which issues during this time. The energy expended upon the motor is then pV. This will be evident from the following considerations:

Proof No. 1.—Suppose a column of water of cross-section o be issuing from the orifice with a velocity v. Since "pressure" means force per unit area, the force driving the water forward is po. This force carries the water forward a distance v in one second; hence the work done per second by the force is pov, and if the experiment last t seconds the total work done is $povt$. But $ovt = V$. Therefore the total energy expended upon the machine during the experiment is pV. Q. E. D.

Proof No. 2.—Suppose the pressure p to be due to a column of liquid of height h, and suppose a mass of M grams of liquid to have issued from the orifice o. In order to restore the conditions existing before the mass M had passed through the orifice, it is necessary to raise M through the height h and to return it to the liquid in the reservoir, i. e., it is necessary to do a quantity of work Mh. This therefore represents the energy which has been expended in the passage of the mass M from the orifice. But if d be the density of the liquid, $p = hd$. $\therefore Mh = M \times \frac{p}{d}$. But $\frac{M}{d} = V$, $\therefore Mh = pV$. Q. E. D.

In this experiment friction is applied to the axle by turning down the thumbscrew s (see Fig. 35) of a Prony brake until the tension in the spring S is just balanced, i. e., until the lever arm cn [$= r$] of the brake rests midway between the stops t, t'. It is generally impossible to eliminate completely oscillation of the lever arm, but its mean position can be estimated with sufficient accuracy. It is evident that the ten-

Work done by motor.

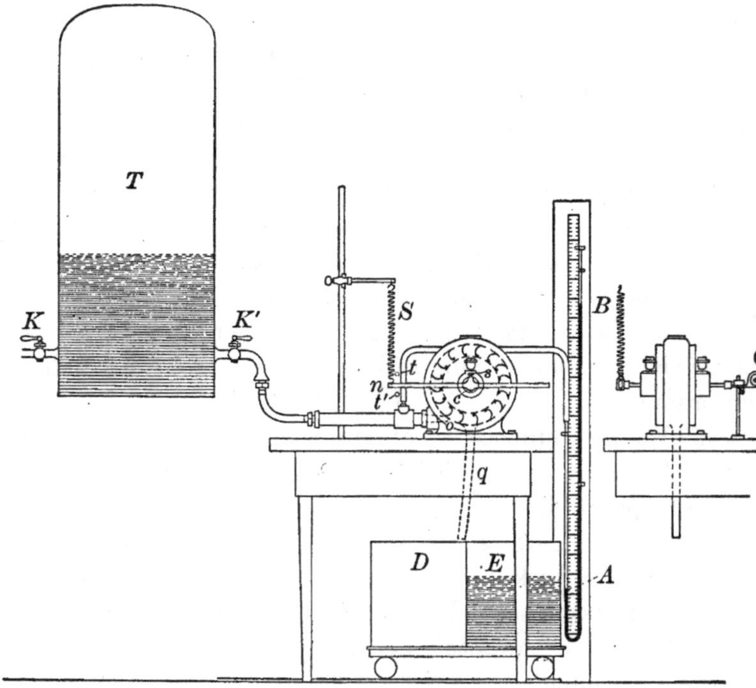

FIGURE 35

sion F in the spring S represents the constant pull which the machine exerts at a distance cn from the axle. If, therefore, cn were the radius of a pulley upon which a cord were being wound up, the constant pull which the cord would exert upon any load which it were moving would be equal to F. One revolution of the water wheel would cause this load to move a distance $2\pi r$. It is evident, then, that the work accomplished by the machine in N

revolutions is $2\pi rNF$. In order to determine the value of F the lower end of the spring is detached from the lever arm, after the run has been made, and known weights added to the spring until it is stretched to the length which it had during the run.

Constant pressure device.

The motor is attached to the regular water supply of the room, but irregularities in the pressure of the latter are equalized by introducing before the motor a large air-tight tank T (200 liters), the entrance and exit both being at the bottom. It is thus the air in the upper part of the tank, compressed by the water-works pressure, which is the immediate source of the pressure applied to the machine.

Pressure measurement.

The pressure under which the water issues from the orifice o is obtained from a reading of the mercury manometer AB. If there were no water in either arm, this pressure, measured in centimeters of mercury, would evidently be the difference between the mercury levels in the two arms of the manometer. This could be reduced to grams by multiplying by the density of mercury. However, since the arm A of the gauge fills with water, and since it is the pressure at the level of o which is sought, the pressure indicated by the mercury height must be diminished by that due to a water column of height equal to the difference between the level of o and the mercury level in A. In solving for efficiency, it is of course essential that p be expressed in the same units as F.

Directions.

Starting with cock K' closed, partially open cock K and allow a considerable pressure to be produced in tank T. Then slowly open K' till the difference in the levels of the mercury in the pressure gauge is, e. g., 100 cm. Then, while one observer holds the gauge-reading constant by continually adjusting K, let another adjust the screw s till the lever arm cn maintains in the mean a horizontal position. These adjustments made, at an accurately observed time deflect the discharge-water, by means of the flexible rubber tube q, from tank E into the empty tank D, and at the same time let a third observer begin to count the revolutions of the speed counter C, which is attached to the axle by means of a flexible rubber connection. When tank D is about two-thirds full, stop the run at an accurately observed time by closing cock K'. Determine V by weighing the water in tank D upon the platform scales. Determine F by stretching the spring

S by means of known weights to the length which it had during the run. If both levels of the mercury in the gauge were not read during the experiment, reëstablish the pressure and read the lower level. Measure the lever arm cn [$= r$], and compute the efficiency. Next vary the tension in the spring S by raising or lowering the support to which the upper end of this spring is attached, and again determine the efficiency for this new "load," the pressure being kept the same as before. In this way make five different runs with loads which vary between zero and the maximum which the machine is able to carry without stopping altogether. The speed will then vary between "racing" speed and the slowest possible speed. Plot a curve with speeds as abscissae and efficiencies as ordinates, and thus determine the speed for which, for the given pressure, the machine is most efficient. If the time of each run be made of exactly the same length the speeds will of course be proportional to the total numbers of revolutions N. In any case, if T represent the *number of minutes* of duration of the experiment and n the speed, $n = \frac{N}{T}$. In this experiment it is recommended that one observer regulate and read pressure, that another attend entirely to the adjustment of the Prony brake (it may need to be continually regulated throughout the run), and that a third observe the time and the number of revolutions of the speed-counter.

Record

1. Wt. of movable block = ——	of W pan = ——	of P pan = ——
1st value of W = ——	of P = ——	∴ efficiency = ——
2d value of W = ——	of P = ——	∴ efficiency = ——
3d value of W = ——	of P = ——	∴ efficiency = ——
4th value of W = ——	of P = ——	∴ efficiency = ——

2. Mean Hg read'g in B —— in A —— Water cor'n —— cm. ∴ p —— gm.

Wt. of D —— of D + water	——	∴ V ——	F ——	T ——	N ——	∴ eff. ——
Second run " "	——	" ——	" ——	" ——	" ——	" ——
Third run " "	——	" ——	" ——	" ——	" ——	" ——
Fourth run " "	——	" ——	" ——	" ——	" ——	" ——
Fifth run " "	——	" ——	" ——	" ——	" ——	" ——

Problems

1. If the friction between two surfaces be proportional to the pressure existing between them, and if, in the experiment with the pulleys, the friction due to the bending of the cord in passing over the pulley be a wholly negligible quantity, how *ought* the efficiency of the system of pulleys to vary with the load?

2. Solve Problems 2 and 3, page 27, from a consideration of the potential and kinetic energies of the body at the top and bottom of the plane and arc.

3. Taking the distance of the sun as 90,000,000 miles, the density of the earth as 5.50, its radius as 4000 miles, find the kinetic energy in kgm.-(meters)^2, which the earth possesses by virtue of its orbital motion. Find how many kilograms of water it could raise from 0° C. to 100° C. if its energy were suddenly to be transformed into heat by a collision. Take 1 kilometer as equal to .62 miles.

Carry to three significant figures only and use throughout the exponential notation, thus 7.14×10^{21}.

4. What mean force has been applied to a kilogram weight if it has been raised 5 meters and at the same time given an upward velocity of 2 meters per second?

5. A bullet enters a target with a velocity of 120 meters per second and penetrates 10 cm. What velocity should it have to penetrate 18 cm.?

6. A Joule is 10^7 ergs. A Watt is an "activity" of 1 Joule per second. A horse-power is an "activity" of 746 Watts. Find the horse-power of a steam pump which lifts 100,000 liters of water per hour from a well 30 meters deep.

7. How many Joules of work are expended by a 200-lb. man in climbing stairs 60 ft. high? How much heat would he develop if he were to fall to the ground, i. e., how many grams of water would the heat generated by the fall raise through 1° C. (1 inch = 2.54 cm., 1 kilo. = 2.2046 lb.)

VI

THE LAWS OF IMPACT

Theory

If two bodies be connected by a stretched or a compressed spring, then if the spring be released, common observation teaches that both bodies are set into motion. According to the Second Law the forces which produce these two motions are measured by the rates of change of momentum of the two bodies. Hence the Third Law in asserting the equality of the two forces asserts the equality at every instant of these two rates of change of momentum. Experiment alone can furnish proof of the correctness of the assertion. But the Third Law is much more than the assertion of this equality in this particular case. It asserts that all forces are essentially like those existing in stretched or compressed springs; in short, that all motions in the universe arise from stresses, and that whenever one body is set in motion some other body always receives the same quantity of motion in the opposite direction. This assertion is a generalization from a large number of experiments upon forces whose effects can be observed and measured. Astronomical observations attest the correctness of the law for gravitational forces. Laboratory experiments must be relied upon to furnish evidence regarding forces arising from impact, magnetization, electrification, etc. In this experiment the law is put to the test for certain cases of impact.

Meaning and proof of Third Law.

When any moving mass m_1 (see Fig. 36) makes impact with any stationary mass m_2, it is observed that the latter is set into motion while the former loses part of its velocity. Consider any instant of the impact and let a_2 be the acceleration which the force f_2 (see Fig.) is

The conservation of momentum.

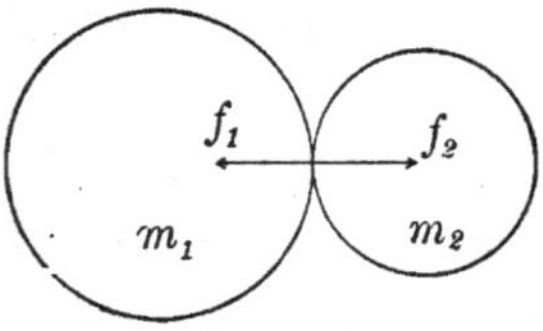

FIGURE 33

imparting at that instant to m_2, and a_1 the negative acceleration which the force f_1 is imparting to m_1. The Second Law gives

$$f_1 = m_1 a_1, \quad \text{and} \quad f_2 = m_2 a_2.$$

The assertion of the Third Law, then, is

$$f_1 = f_2, \quad \text{or} \quad m_1 a_1 = m_2 a_2.$$

But if the rates of change of momentum ot m_1 and m_2 are at every instant equal, it follows that the *total* momentum imparted to m_1 during the whole time of impact is equal to the *total* (opposite) momentum imparted to m_2; or symbolically, if u_1 represent the velocity of m_1 before impact, v_1 its velocity after impact, and v_2 the velocity imparted to m_2,

$$m_1(u_1 - v_1) = m_2 v_2;$$

or by transposition

$$m_1 u_1 = m_1 v_1 + m_2 v_2, \tag{32}$$

an equation which asserts that *the total momentum before impact is equal to the total momentum after impact.*

If m_2 has an initial velocity u_2, precisely the same line of reasoning gives

$$m_1 u_1 + m_2 u_2 = m_1 v_1 + m_2 v_2. \tag{33}$$

Thus the Third Law asserts that *momentum is conserved in all impacts, be it between elastic or inelastic bodies.*

Energy relations after impact.

While thus the Third Law asserts that there is never any loss of momentum in an impact, it does *not* assert that there is no loss in kinetic energy. The mechanical energy is always less after impact than before, and in the case of an inelastic impact between two bodies one of which is at rest ($u_2 = 0$) the loss can be theoretically calculated from a knowledge of the masses alone. Thus, for this case, since in inelastic impact the bodies remain together after the collision, $v_1 = v_2$, and the general equation (33) becomes

$$m_1 u_1 = (m_1 + m_2) v_2. \tag{34}$$

Substituting the value of v_2 obtained from (34) in the expression which represents the fractional loss l of kinetic energy, viz.,

$$l = \frac{K\,E \text{ before} - K\,E \text{ after}}{K\,E \text{ before}} = \frac{\frac{1}{2} m_1 u_1^2 - \frac{1}{2}(m_1 + m_2) v_2^2}{\frac{1}{2} m_1 u_1^2}, \tag{35}$$

there results, after reduction, the simple formula,

$$l = \frac{m_2}{m_1 + m_2}. \tag{36}$$

The Third Law therefore leads to the interesting conclusion that the per cent loss in kinetic energy, when one inelastic body impinges upon another which is at rest, is altogether independent of the velocity of the impinging body.

Experiment

Object. To test the equality of momenta before and after impact for the case of inelastic impact.

It is not easy to measure *directly* the velocities immediately before and after an impact. But if the velocity of the impinging body be acquired by a fall from a known height h_1, and if the struck body expend its velocity in rising to a known height h_2, the required velocities u_1 and v_2 can be easily calculated from the measured heights h_1 and h_2. *Method.*

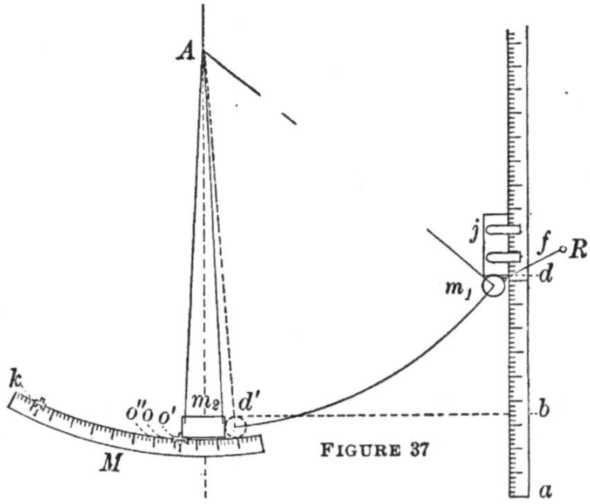

FIGURE 37

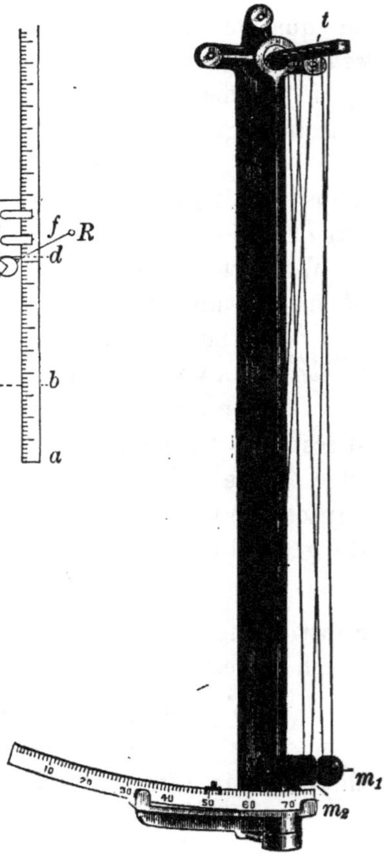

FIGURE 37a

In order to realize these conditions a lead ball m_1, rendered perfectly inelastic by means of a piece of soft wax, is allowed to fall down a circular arc from the position d to the position d' (see Fig. 37). In so doing it acquires the same velocity as though it had fallen through the vertical distance db (see Ex. III, Problem **3**, and Ex. V, Problem **2**). After impact the cylinder and ball move together up the arc M, which is graduated in degrees, to a point which is registered by the light aluminum index 1. The vertical height corresponding to this movement up the arc is calculated from the index readings as follows:

Calculation of h_2 from index readings.

Let o be the index reading when ball and cylinder hang together. The center of gravity of the system formed by the ball and cylinder together must then be at some point c (see ideal diagram, Fig. 38) which is directly underneath the point of support A. Let o' be the index reading when the cylinder hangs alone, i. e., the reading corresponding to the position of the system at the instant of impact. At this instant the center of gravity of the system is at some point c', which is to the right of c a distance such that $c - c' = o - o'$. Now if the force of the blow were zero, as soon as the ball and cylinder were joined, the force of gravity alone would cause the system to move forward, for c, not c', is the natural position of its center of gravity. The velocity acquired in falling from c' down to c would of course carry the center of gravity of the system up again to some point c'' such that $c - c' = c'' - c$. It is evident, then, that only the movement over the arc $c''c'''$ [$= o''k$] can be attributed to the velocity imparted by the blow. Hence h_2 is the vertical height corresponding to this portion only of the arc, i. e. (see Fig. 38):

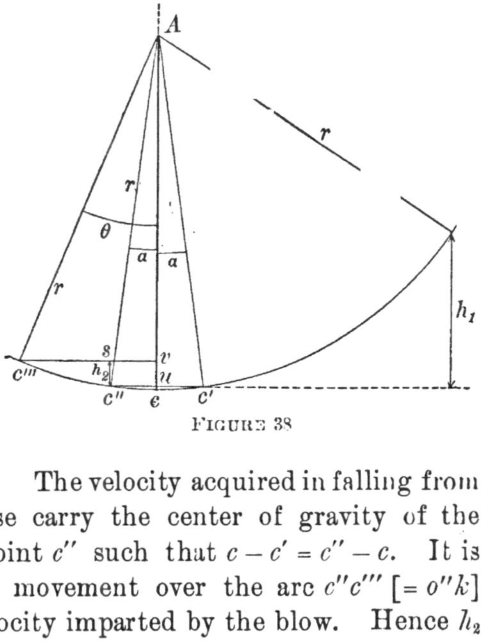

FIGURE 38

$$h_2 = c'' s. \tag{37}$$

But
$$c''s = uv, \quad \text{and } uv = uA - vA. \tag{38}$$

Now
$$\frac{uA}{r} = \cos a, \text{ and } \frac{vA}{r} = \cos \theta. \tag{39}$$

Hence
$$uA - vA = r(\cos a - \cos \theta), \tag{40}$$

in which θ is the angle whose arc is cc''' [$= ok$], and a is the angle whose arc is $c''c$ [$= cc' = oo'$]. Since the arc is graduated in degrees,

a and θ are obtained at once from the readings. Finally, then, from (37), (38), and (40),

$$h_2 = r\ (\cos a - \cos\theta). \tag{41}$$

The radius r should be measured from the point of support to the common center of gravity. Cos a and cos θ may be obtained from any trigonometrical table.

The equations to be verified.

The momentum equation which it is sought to verify is $m_1u_1 = (m_1+m_2)v_2$. But, since $u_1 = \sqrt{2gh_1}$ and $v_2 = \sqrt{2gh_2}$, this equation may be written in the form

$$m_1\sqrt{h_1} = (m_1+m_2)\sqrt{h_2}. \tag{42}$$

By a similar substitution the expression for the loss of kinetic energy (35) may be written

$$l = \frac{m_1h_1 - (m_1+m_2)h_2}{m_1h_1}. \tag{43}$$

Directions.

Make several preliminary trials, adjusting, if need be, the positions of the clamp R (Fig. 37), and the suspensions of the ball and cylinder by means of the thumbscrews t (Fig. 37a), until, when the ball is released by burning the thread f, the cylinder moves smoothly up the arc without wabbling. This done, measure the height from the top of the table to the top of the ball by means of a meter stick furnished with a sliding clip j (Fig. 37). Then move the index up the arc to nearly the point which will be reached by the cylinder. After the impact take very carefully the index readings at k, o, and o', and also the distance from the table or floor to the top of the ball when the latter is held at the point of impact. (It must of course be provided that the reference plane from which the heights ab and ad are measured is accurately horizontal.)

Obtain at least two sets of readings. Weigh the ball and cylinder upon the trip scales.

Record

1st Trial			2d Trial		
m_1 = ——	m_2 = ——		m_1 = ——	m_2 = ——	
ad (Fig. 37) ——	ab ——	$\therefore h_1$ ——	ad = ——	ab = ——	$\therefore h_1$ = ——
Read'g at k ——	at o ——	at o' ——	at k = ——	at o = ——	at o' = ——
r = ——	θ = ——	a = ——	r = ——	θ = ——	a = ——
Mom. bef. ——	aft. ——	% dif. ——	Mom. bef. ——	aft. ——	% dif. ——
l from (43) ——		from (36) ——	l from (43) = ——		from (36) = ——

Problems

1. What relation exists between m_1 and m_2 when $\frac{1}{2}$ of the K E is lost in heat? When $\frac{1}{4}$? When $\frac{3}{4}$?

2. The moon moves toward the earth a distance of 15 ft. per minute. Find how far the earth moves toward the moon in the same time. (Ratio of masses, 1 to 81.4.)

3. A rifle bullet weighing 20 gm. was fired into a ballistic pendulum weighing 4 kilos. The latter moved up an arc a distance corresponding to a vertical rise of 5 cm. Find the velocity of the bullet.

4. Is it true that, at the start, the wagon pulls back with the same force with which the horse pulls forward? If so, how is any motion produced? If not, reconcile your answer with the Third Law.

5. A bullet weighing 20 gm. and having a speed of 300 meters per sec., struck and imbedded itself in a bird weighing 5 kilos. which was flying in the same direction as the bullet with a speed of 150 kilometers per hour. Find the velocity of the bird, in kilometers per hour, the instant after it was shot.

6. What becomes of the momentum of a meteorite which collides with the earth? What becomes of its energy?

7. Two equal inelastic balls moving with equal velocities in opposite directions collide. Show that in this case the momentum before impact is the same as that after impact.

Note that velocity is a directed quantity.

8. A billiard ball weighing 100 gm. and moving east with a speed of 2 meters per sec. was struck by a putty ball weighing 4 gm. and moving south with a speed of 20 meters per sec. Find the speed and direction of the ball the instant after the impact.

9. A 500 gm. bird sat on a pole 30 meters high. A boy standing 20 meters from the base of the pole shot the bird with a 10 gm. bullet which had a speed of 150 meters per sec. How far did the bird rise above the pole? How far from the base of the pole did it strike the ground?

Assume that the bullet lodged in the bird.

VII

ELASTIC IMPACT. COEFFICIENT OF RESTITUTION

Theory

Since force is equal to rate of change of momentum it follows from the Third Law that the mean force acting between two impinging bodies is the total change in the momentum of either divided by the time of duration of the impact. Since this time can not in general be determined, it is customary to confine attention to the total change in momentum which each body experiences by virtue of the impact. This quantity is called the "impulse" of the force, and will be hereafter represented by the letter R. If, then, u_1 and v_1 represent the velocities of the first body m_1 before and after impact respectively, u_2 and v_2 the velocities of the second body m_2 before and after the impact, then by definition

Definition of impulse.

$$R = m_1 (u_1 - v_1), \tag{44}$$

or (see Third Law),

$$R = m_2 (v_2 - u_2). \tag{45}$$

In the case of elastic impact, it is convenient to divide R into two parts R_1 and R_2, of which R_1 represents the impulse *during the compression*, and R_2 the impulse *from the instant of greatest compression to the instant of separation.*

In impacts between *inelastic* bodies $R_2 = 0$. If the colliding bodies are perfectly elastic, it might be expected that R_2 would be equal to R_1. In point of fact this is never the case, for there are losses due to internal friction even with bodies which when subjected to static tests show perfect elasticity (see definition of perfect elasticity in Ex. VIII). However Newton proved experimentally that for any two given bodies the ratio $\frac{R_2}{R_1}$ is always a constant so long as the impact is not so violent as to produce permanent deformation.

Definition of coefficient of restitution.

This ratio is called the *coefficient of restitution*, and will be hereafter represented by the letter e. It is always less than unity.

This coefficient of restitution $e\left[=\frac{R_2}{R_1}\right]$ may also be shown to be the velocity of recession, $v_2 - v_1$, of the two colliding bodies divided by their velocity of approach, $u_1 - u_2$. For it is evident that at the instant of greatest deformation m_1 and m_2 have a common velocity. Call this velocity S. Then, from the above definitions of R_1 and R_2,

Proof that $e=\frac{v_2-v_1}{u_1-u_2}$.

$$R_1 = m_1\ (u_1 - S) = m_2\ (S - u_2), \tag{46}$$

$$R_2 = m_1\ (S - v_1) = m_2\ (v_2 - S). \tag{47}$$

Division of (47) by (46) gives

$$\frac{R_2}{R_1}\ [= e] = \frac{S - v_1}{u_1 - S} = \frac{v_2 - S}{S - u_2}. \tag{48}$$

From (48) come the two equations,

$$S - v_1 = e\ (u_1 - S), \tag{49}$$

$$v_2 - S = e\ (S - u_2). \tag{50}$$

Addition of (49) and (50) gives

$$v_2 - v_1 = e\ (u_1 - u_2). \tag{51}$$

Q. E. D.

In the special case of impact against a fixed plane, $u_2 = 0$, $v_2 = 0$, and v_1 is opposite in direction to u_1 therefore

$$e = \frac{v_1}{u_1}. \tag{52}$$

Newton's law as to the constancy of e may therefore be very easily tested by varying the velocity of approach of a body toward a fixed plane and measuring the corresponding velocities of rebound. If the body be dropped vertically upon the fixed plane, the velocities u_1 and v_1 can easily be determined from the heights of fall and of rebound.

The actual loss L of mechanical energy upon impact (not the fractional loss l) is evidently

Loss of kinetic energy in impact.

$$L = (\tfrac{1}{2}m_1u_1^2 + \tfrac{1}{2}m_2u_2^2) - (\tfrac{1}{2}m_1v_1^2 + \tfrac{1}{2}m_2v_2^2); \tag{53}$$

or

$$L = \tfrac{1}{2}m_1\ (u_1 - v_1)\ (u_1 + v_1) + \tfrac{1}{2}m_2\ (u_2 - v_2)\ (u_2 + v_2); \tag{54}$$

or from (44) and (45)

$$L = \frac{R}{2}(u_1 + v_1) - \frac{R}{2}(u_2 + v_2) = \frac{R}{2}\{(u_1 - u_2) - (v_2 - v_1)\}. \quad (55)$$

The combination of (55) with (51) gives

$$L = \frac{R}{2}(u_1 - u_2)(1 - e). \quad (56)$$

But substitution in (51) of the values of v_2 and v_1 obtained from (44) and (45) gives

$$R = \frac{m_1 m_2}{m_1 + m_2}(1 + e)(u_1 - u_2). \quad (57)$$

Substitution of this value in (56) gives for the loss of kinetic energy

$$L = \tfrac{1}{2}(1 - e^2)(u_1 - u_2)^2 \frac{m_1 m_2}{m_1 + m_2}. \quad (58)$$

The fractional loss l is this expression divided by the initial energy. Hence for the simple case in which m_2 is at rest, i.e., for which $u_2 = 0$

$$l = (1 - e^2)\frac{m_2}{m_1 + m_2}, \quad (59)$$

and for the still simpler case in which m_2 is not only at rest but is also infinitely large, i.e., the case of impact upon a fixed plane,

$$l = (1 - e^2). \quad (60)$$

Equation (59) shows that, as in the case of inelastic impact, the loss in kinetic energy is independent of the velocity of the striking body provided the struck body is at rest. When $e = 0$ (59) reduces to (36).

It appears from (58) that the sole condition of no loss of kinetic energy in an impact is $e = 1$. The fact that e is always somewhat less than unity means then that in all impacts there is some transformation of mechanical into heat energy.

Velocities after impact.

It is evident from the general momentum equation (33), viz.,

$$m_1 u_1 + m_2 u_2 = m_1 v_1 + m_2 v_2,$$

and from the equation (51) which determines e, viz.,

$$e = \frac{v_2 - v_1}{u_1 - u_2}, \quad (61)$$

that the velocities after impact can always be found if the coefficient of restitution, the two masses, and the two initial velocities are known. For the simple case in which $e = 1$, i. e., for the case of so-called "perfectly elastic" * impact, the solution of these two equations gives

$$v_1 = \frac{2m_2u_2 + u_1(m_1 - m_2)}{m_1 + m_2}, \tag{62}$$

$$v_2 = \frac{2m_1u_1 - u_2(m_1 - m_2)}{m_1 + m_2}. \tag{63}$$

Experiment

Object. **1.** To test Newton's law as to the constancy of e; **2.** to prove the equality of momenta before and after impact and to find the coefficient of restitution and the per cent loss of mechanical energy in the impact of two steel balls.

Directions. **1.** Drop steel and glass balls from the clamp c (see Fig. 39) through the ring o, and slip the horizontal rod r up or down the vertical support, until, in the rebound from the smooth top of the heavy steel plate P†, the bottom of the ball just becomes visible above o. Make observations for at least three different heights of fall which lie between, say, 30 cm. and 100 cm. In each case make the measurement from the steel plate to the *bottom* of the ball.

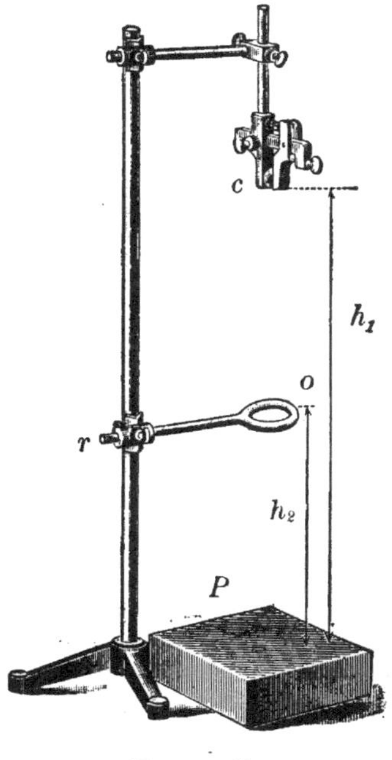

FIGURE 39

Since in general for falling bodies starting from rest $v = \sqrt{2gh}$,

$$e = \frac{v_1}{u_1} = \sqrt{\frac{h_2}{h_1}}. \tag{64}$$

2. The form of apparatus used in **2** (see Fig. 40) is the same as that used in Ex. VI, save that all three heights, viz., the height of *fall* of m_1

*The term is rather unfortunate since "perfect elasticity" is often used in a somewhat different sense (see Ex. VIII).

† A slab of slate or any smooth, hard, and heavy body may replace the steel plate.

before impact and the heights of *rise* of both m_1 and m_2 after impact are measured upon a graduated arc.

Apparatus.

Directions.

First make such adjustment of the lengths of the supporting strings that, when the balls hang freely, the wire frame carried upon the bottom of m_2 clears index 1 (see Fig. 41), but catches index 2, while the like frame on m_1 catches index 1. Then adjust so that when the thread which holds m_1 back is burned, m_2 is driven straight up the arc by the impact from m_1. Take the reading of index 1 when m_1 is tied in position a. Next slide index 1 down nearly to b, the point to which m_1 will move after the impact (this point should be approximately located by a preliminary trial), and place index 2 in the neighborhood of c. Then burn the thread, catch m_1 with the hand as it swings back after the impact, and take the readings at b and c. Note the zero reading of each ball, i. e., the reading when each ball hangs *alone*, and finally take the reading of index 1 when m_1 is at the point

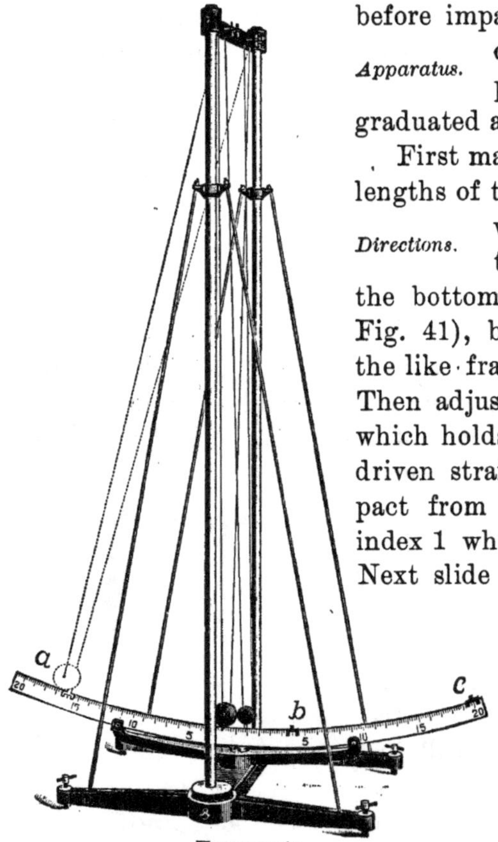

FIGURE 40

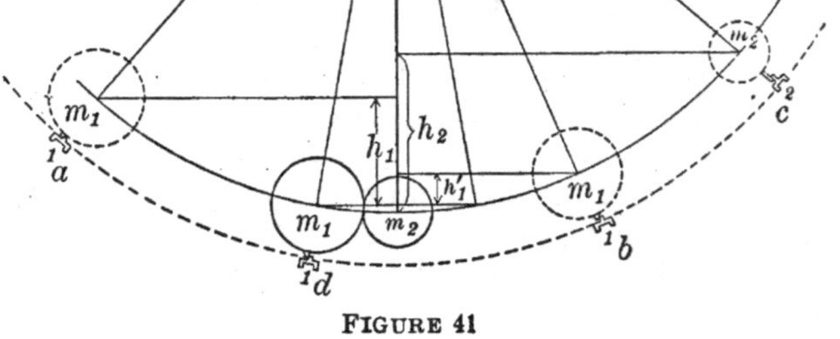

FIGURE 41

of impact d. *This is not the reading when both balls hang together*, but the reading when m_2 hangs freely and m_1 is brought down so as just to touch m_2.

Calculations. If α represent the difference between the zero reading of index 1 and the reading at the point of impact d, θ the difference between the zero of 1 and the reading at a, β the difference between the zero of 1 and the reading at b, and ω the difference between the zero of index 2 and the reading at c, it is evident from the discussion of Ex. VI that the height h_1 through which m_1 falls before impact, the height h_2 through which m_2 is raised by the impact, and the height h_1' through which m_1 rises by virtue of the velocity which it retains after the impact, are given by the relations (see Fig. 41)

$$\left.\begin{aligned} h_1 &= r(\cos\alpha - \cos\theta), \\ h_2 &= r(1 - \cos\omega), \\ h_1' &= r(\cos\alpha - \cos\beta). \end{aligned}\right\} \qquad (65)$$

The equation to be verified, viz.,

$$m_1u_1 = m_2v_2 + m_1v_1,$$

may be written, since $v = \sqrt{2gh}$,

$$m_1\sqrt{h_1} = m_2\sqrt{h_2} + m_1\sqrt{h_1'}; \qquad (66)$$

or

$$m_1\sqrt{(\cos\alpha - \cos\theta)} = m_2\sqrt{(1 - \cos\omega)} + m_1\sqrt{(\cos\alpha - \cos\beta)}. \qquad (67)$$

A similar substitution in the expression for the coefficient of restitution [see (61)] gives

$$e = \frac{\sqrt{(1-\cos\omega)} - \sqrt{(\cos\alpha - \cos\beta)}}{\sqrt{(\cos\alpha - \cos\theta)}}. \qquad (68)$$

The loss of energy is to be obtained from e, m_1 and m_2 [see (59)].

Record

	Steel			Glass		
1.—Ht. of fall	1st ——	2d ——	3d ——	1st ——	2d ——	3d ——
Ht. of rebound	" ——	" ——	" ——	" ——	" ——	" ——
∴ e =	" ——	" ——	" ——	" ——	" ——	" ——
% loss of K E	" ——	" ——	" ——	" ——	" ——	" ——

2.—Rdg. of index 1 at a —— at b —— at zero —— at pt. of impact ——
" " " 2 " —— at c ——

m_1 —— m_2 —— ∴ α = —— ∴ θ = —— ∴ β = —— ∴ ω = ——

Momentum before —— after —— e = —— % loss of K E = ——

Problems

1. Show from (57), (44), and (45) that when two equal balls, for which $e = 1$, collide centrally, they simply exchange velocities, i. e., the result is the same as though one had passed through the other without in any way influencing it.

2. Hence explain why the only effect of a central impact of one marble upon a row of marbles is to drive off the end marble. Also why the impact of two marbles drives off two from the end, etc.

3. A 300 gm. ball approaches a bat with a velocity of 50 meters per second; it leaves with an opposite velocity of 100 meters. Find the mean force of the blow if the impact last $\frac{1}{50}$ second.

4. A rapid-fire gun shoots 500 30-gm. bullets per minute. Find what force is necessary to hold it in place if the velocity of the bullets be 500 meters per second.

For a constant force, or for a succession of impulses so rapid that the effect is the same as that of a constant force, "rate of change of momentum" is change of momentum per second.

5. A fire engine throws 400 liters of water per minute from a pipe furnished with a nozzle of 4 cm. diameter. What force does a wall experience against which the jet is directed at short range (assume inelastic impact)? If each particle of water were "perfectly elastic," i. e., rebounded with the velocity of approach, what would be the value of the force?

6. A bullet weighing 50 gm. is fired into a block weighing 125 gm. Find the per cent loss of mechanical energy. Had ball and block been elastic bodies for which $e = 1$, what would have been the loss?

7. e is very nearly unity for equal spheres made of perfectly elastic materials (steel, glass, ivory, etc.), but it is not unity for unequal balls of the same substances. It may be as low as .75 for steel balls of greatly different size. Thus e is a constant of the colliding bodies, *not* of the material. Why?

Consider vibration losses, and the conditions under which vibrations will persist in the bodies after impact.

8. Find the velocities after impact of two directly impinging bodies whose masses are 50 gm. and 100 gm., whose velocities before impact are in the same direction and have values of 600 cm. and 350 cm. respectively, and for which e is unity. Ditto for balls for which $e = .90$.

9. Explain the rise of a rocket.

VIII

ELASTICITY

HOOKE'S LAW: YOUNG'S MODULUS

Theory

Most substances possess in greater or less degree two quite distinct kinds of elasticity: (1) elasticity of volume, (2) elasticity of form. A body is said to have *volume elasticity* if it tends to return to its original volume after being compressed or dilated by the application of force; i. e., if its molecules tend to maintain fixed *distances* with reference to one another, and resist any attempt to increase or decrease these distances. A body possesses form elasticity, or *rigidity* if its molecules tend to maintain a fixed *configuration*, and resist any attempt to produce *slipping* motions among themselves.

Two kinds of elasticity.

The "*volume coefficient*" or the "*volume modulus*" of elasticity is a constant which measures the restoring force called into play by a given change in the mean distance between adjoining molecules, the configuration remaining unchanged.

The two coefficients.

The *coefficient of rigidity* or "*rigidity modulus*" is the constant which measures the restoring force called out by a given change in the relative positions of the molecules, the mean distance remaining unchanged. No connection whatever exists between the two kinds of elasticity. Thus all liquids possess a volume modulus which is enormous, a rigidity modulus which is essentially zero. India-rubber has nearly the same volume modulus as water, but with a very pronounced rigidity modulus. The metals have very large moduluses of both volume and form.

A body is said to be *perfectly elastic* if it always requires the same force to produce the same displacement. Thus a wire would show perfect elasticity if the successive removal of a number of stretching weights caused it to resume exactly the lengths which it had during the successive addition of the weights, and that no matter how long or how

Perfect elasticity.

short a time the weights had been in place. If *static* experiments only are taken as the test of perfect elasticity, all liquids are perfectly elastic, and most solids also so long as the displacements are kept within certain limits. The limits of perfect elasticity, however, differ very widely for different substances. Jellies and rubber show perfect *rigidity* through very wide limits, iron through very small, lead through smaller still, etc.

Important distinctions.

There is no connection between the "degree" of elasticity of a body and its elastic constants. The former measures the perfectness of the return to the initial condition, the latter the magnitude of the force required to produce a given displacement. Thus jelly is nearly perfectly rigid, but has a very small rigidity coefficient. Lead has large coefficients, but a small degree of elasticity. In popular usage a body is said to be "very elastic" which possesses nearly perfect elasticity through wide limits. Properly speaking, ability to rebound is a measure of resilience, not of "degree of elasticity." Nevertheless, the former depends in large measure upon the latter.*

Hooke's Law.

Experiment shows that within the limits of perfect elasticity the *displacement* produced is *proportional to the force applied*, whether it be the bulk or the form elasticity which is made the subject of test.

The volume modulus.

For the determination of the volume modulus, force must be applied in such a way that it will compress or dilate the body equally in all directions, but will not distort it. This can of course be done only by a uniform pressure or tension applied to all points of the surface of the body. The bulk modulus k is then defined as *the ratio between the force per unit area* (the stress) *and the change in volume per unit volume* (the strain). By Hooke's Law this ratio must be constant. Thus,

$$\text{vol. mod.} = \left(\frac{\text{force}}{\text{area}}\right) \div \left(\frac{\text{change in vol.}}{\text{vol.}}\right),$$

or symbolically, since by definition $\frac{\text{force}}{\text{area}} = \text{pressure}$,

$$k = \frac{P}{\frac{v}{V}} = \frac{PV}{v} \tag{69}$$

* Art. on "Elasticity" by Lord Kelvin in *Encyclopædia Britannica.*

in which k stands for the volume modulus, P for pressure (i. e., force per unit area), V for volume, and v for change in volume.

Unfortunately the direct determination of k is not easy. It is therefore customary to determine instead a quantity called

Young's modulus.

Young's modulus, which is a cross between the volume modulus and the rigidity modulus. A known weight is hung from a wire and the elongation measured. *Young's modulus is then defined as the ratio between the force per unit cross-section of the wire and the elongation per unit length.* It is evident that the operation here described tends to produce a change in form as well as in volume, for instead of increasing or decreasing all dimensions it increases the length and decreases the diameter. Analysis which is beyond the scope of this book shows that from this hybrid modulus Y, and the rigidity modulus n (see Ex. IX) the bulk modulus k can be found by the equation

$$Y = \frac{9nk}{3k + n}. \tag{70}$$

Experiment

Object.

To test Hooke's Law and to find Young's modulus for steel.

The instrument for determining the elongation produced in a wire by a given stretching force is shown in Fig. 42. The upper end of the wire is

Description.

firmly clamped in a chuck at a. At b the wire is gripped by a second chuck which is set into a cylindrical brass piece. This cylinder passes, with very little play, through a circular hole in the cross piece n. n is rigidly clamped

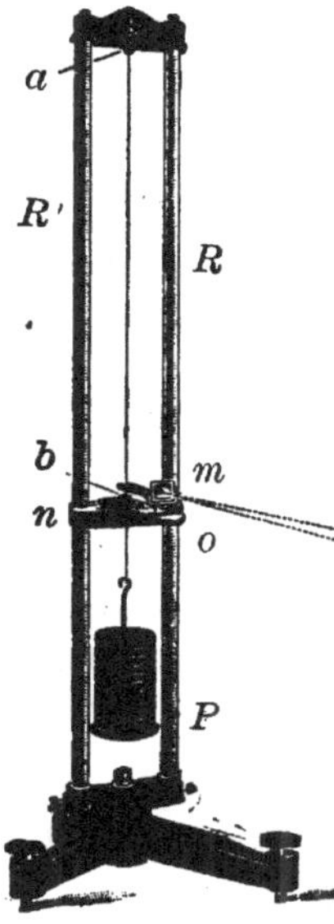

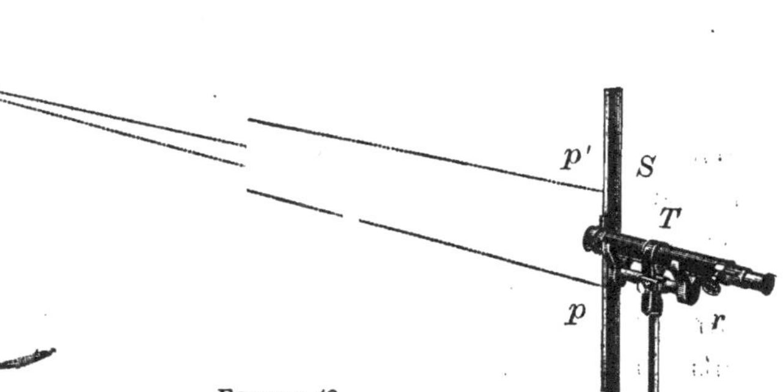

FIGURE 42

to the upright rods R and R', and carries a small horizontal table upon which rest the front feet of the optical lever m. In order to prevent easy displacement of the lever these feet are set in a groove o. The rear foot of the lever rests upon the face of the chuck b. At a distance of twelve or thirteen feet from the instrument a telescope T, to which is attached a vertical scale S, is so placed that the image of the scale formed by the mirror m is visible in the telescope. The addition of a weight to P stretches the wire and lowers the rear foot of the lever a distance which will be denoted by ϵ. If l represent the distance from the rear foot of the optical lever to the mid-point between its front feet, it is evident that the production of the elongation ϵ in the wire ab has caused the mirror m to turn through an angle of $\frac{\epsilon}{l}$ radians. This angular motion of the mirror causes some point p' instead of p to come under the cross-hairs of the telescope. The beam of light reflected by the mirror has thus been turned through the angle $\frac{pp'}{mp}$ radians. But this angle is twice that through which the mirror turns (see Problem 1 below); hence the distance ϵ is easily determinable in terms of the measurable quantities pp', mp, and l.

Directions.—First locate the image of the scale in the telescope. To do this move the head about near the telescope until the image of the eye is seen in the mirror m. If the eye cannot be found at the distance of the telescope, move the head up toward the mirror until it *is* found; then move back again. Keeping the image of the eye in sight, move the telescope and scale into, or near to, the position occupied by the eye. Then adjust positions until, when the eye sights *over*, *not through*, the telescope, the image of the scale is seen in the mirror. The image of the scale must then fall upon the objective of the telescope. Then looking through the telescope, focus by means of the rack and pinion r until the mirror itself is seen, then slowly push *in* the eye-piece by means of r until the scale is brought into view. If the cross-hairs are not in sharp focus, move the eye-piece alone, in or out, until they appear perfectly sharp, then refocus upon the scale by means of r.

Finding the scale in the telescope.

Next turn the mirror m until the portion of the scale seen in the telescope is that near the objective, and take a careful reading of the position of the cross-hairs upon the scale, estimating to tenths millimeters. If a slight change in the position of the eye changes at all the reading, focus again carefully until this "error of parallax" is altogether removed. Add kgm. weights successively to the pan and take the corresponding readings. Make similar records as the weights are successively removed. If upon removal of the weights the cross-hairs do not return to their original position, the limit of perfect elasticity has been overstepped and the readings must be repeated with the use of fewer weights. In adding or removing weights, use extreme care to prevent jarring the instrument or changing the position of the lever-point on chuck. A divergence in successive elongations of more than one per cent indicates carelessness.

The scale readings.

In order to determine l take the imprint of the three feet of the optical lever upon a sheet of paper. With a knife-edge draw a line connecting the centers of the two front feet. The distance from the middle point of this line to the center of the third foot may be measured with a steel rule held on edge. Estimate to tenths millimeters. Measure the distance mp with a tape.

The other measurements.

Since in this case Young's modulus involves the sectional area of a very small wire, it is necessary that the diameter be measured with great care. Measure with micrometer calipers, and let the final result be a mean of at least a dozen observations taken at equal intervals from top to bottom of the wire. In using the calipers always take the zero reading as well as the reading when the wire is between the jaws.

In the calculation of Young's modulus, the *stress* (force per unit cross-section) must be expressed in dynes per square centimeter, the *strain* (elongation per unit length) in centimeters per centimeter. It is also required to plot a curve in which the total weights in the pan at each addition shall represent abscissae, and the corresponding total elongations, measured from the first reading, shall represent ordinates. As a check upon the accuracy of the work, change the position of the telescope and scale, and make a second complete determination of Y.

The use of the measurements.

Record

	1st Determination			2d Determination		
Wts.	Rdgs.	Difs.	Diams.	Rdgs.	Difs.	Diams.
1.						
2.						
3.						
4.						
5.						
6.						
7.						
6.						
5.						
4.						
3.						
2.						
1.						
	Means					

$l =$ —— $mp =$ —— wire length = —— $l =$ —— $mp =$ —— $\therefore \epsilon =$ ——

$\therefore \epsilon =$ —— ..... $\therefore Y =$ —— $\times 10^{11}$ $\therefore Y =$ —— $\times 10^{11}$. % dif. = ——

Problems

1. From the optical law *angle of incidence equals angle of reflection*, prove that a beam of light reflected by a mirror turns through twice the angle through which the mirror turns.

2. Can a body whose bulk modulus is infinite have a finite Young's modulus?

3. A wire 80 cm. long and .3 cm. in diameter is stretched .3 mm. by a force of 2 kilo. How much force would be required to stretch a wire of 180 cm. length and 8 mm. diameter through 1 mm.?

4. An iron and a brass wire have each the length of 15 cm. when each is stretched by a force of 1 kgm. The length of the iron wire becomes 15.4 cm. under a stress of 3 kgm. and that of the brass wire becomes 15.6 cm. under a stress of 9 kgm. Compare the modulus of iron with that of brass.

5. What force is needed to double the length of a steel rod whose diam. is 2 mm.? Assume perfect elasticity.

IX

THE COEFFICIENT OF RIGIDITY

Theory

Definition and illustration of shear.

In order to find the *coefficient of rigidity* of a substance it is necessary to apply a force which will cause the molecules to shift their *relative positions* without altering at all their distances apart. Such a change is called a "shear." To take as simple a case as possible, imagine a rigid cylindrical shell whose wall is one molecule in thickness, and let the height of the cylinder be so small that it contains but two rows of molecules [see Fig. 43 (1)]. Let a tangential force act upon each of the molecules 1, 2, 3, 4, etc., of the upper row so as to bring them into the positions shown in Fig. 43 (2), the molecules of the lower row being held fast by equal and opposite forces. The change produced is evidently a pure shear, since *configuration* alone has been changed, all *distances* remaining exactly as at first. The *shearing force* is the total force which has acted, or the sum of the forces upon the individual molecules. The angle through which the line connecting any two molecules which originally lay in the same vertical line has been turned by the shearing force, is taken as the *measure of the shear.* This angle θ is always expressed in radians.

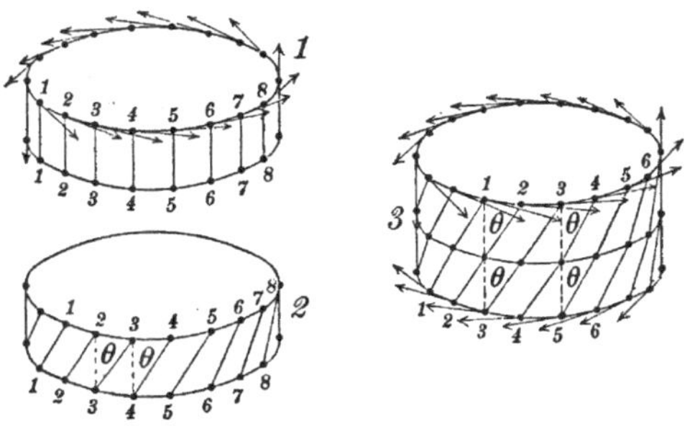

FIGURE 43

Were the cylinder three rows of molecules in height instead of two [see Fig. 43 (3)], then, upon the application of the same shearing force, the upper row would move twice as far as before, but the angle of shear θ would remain the same, for the case would be precisely the same as though the middle row were clamped fast and the equal and opposite forces on the upper and lower rows produced each the same effects which were considered in case 1. From an extension of the same line of reasoning to a still longer cylinder, it is evident that the shear produced by the application of a given shearing force to the top of the cylinder is independent of the height of the cylinder, for the shearing motion ultimately ceases only when the restoring force due to the rigid connection between the top row and the next to the top row of molecules is equal to the shearing force, i. e., when there is a given angular displacement between these two rows. In order that the *second* row from the top may be in equilibrium, *the same* angular displacement must exist between it and the third row as exists between rows 1 and 2, and so on to the bottom row. Thus a given shearing force must produce a given angular tilt in a row of vertical molecules whether the cylinder be long or short.

Shear independent of height of cylinder.

Conceive now the ideal cylindrical shell to be replaced by an actual thin hollow cylinder of length l, of mean radius r and of thickness t [see Fig. 44 (1)]. Divide the upper surface into unit areas, and let a tangential force f be applied to each unit area [see Fig. 44 (2)]. *The coefficient of rigidity n is now defined as the ratio between the force per unit area*

Definition of coefficient of rigidity n.

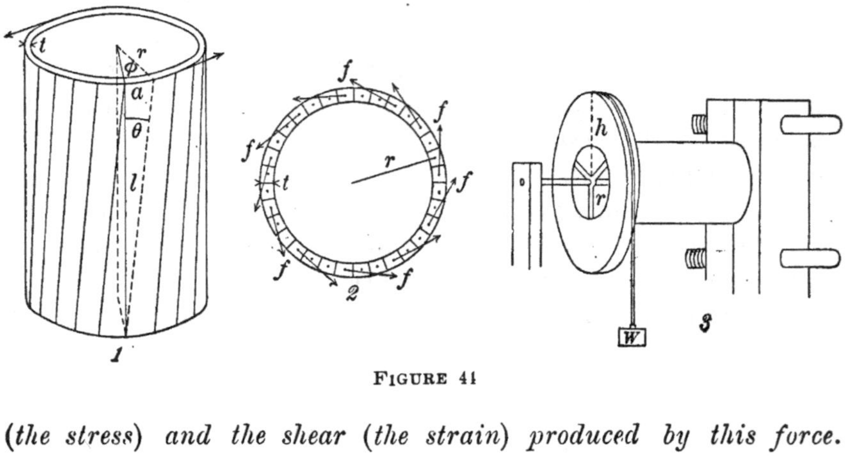

FIGURE 44

(the stress) and the shear (the strain) produced by this force. Symbolically [see Fig. 44 (1)],

$$n = \frac{f}{\theta}. \tag{71}$$

The total shearing force is, however, Af. Call this force f' and the area of the ring A. Then $n = \frac{\text{shearing force}}{\text{area}} \div$ angular displacement; or

$$n = \frac{f'}{A\theta}. \tag{72}$$

From Hooke's Law this ratio is the same for all values of f'.

It is not so easy to measure θ directly as it is to measure ϕ [Fig. 44 (1)], the angle through which the end of the cylinder is twisted by the force f'. Since the arc a [Fig. 44 (1)] is always small in comparison with the length of the cylinder, it is possible to write without appreciable error,

Measurement of n from hollow cylinder.

$$\frac{a}{l} = \theta. \qquad \text{But also} \qquad \frac{a}{r} = \phi.$$

Hence

$$\theta = \frac{\phi r}{l}. \tag{73}$$

Again

$$A = 2\pi rt. \tag{74}$$

Substitution of these values of A and θ in (72) gives

$$n = \frac{f'l}{2\pi r^2 t\phi}. \tag{75}$$

If the force f' be called into play in the manner shown in Fig. 44 (3), i. e., by clamping one end of the cylinder, screwing the other rigidly to a grooved circular disk and applying a twisting force of say F [$= Wg$] dynes to the circumference of the disk by means of weights W, then by the principle of moments [see Fig. 44 (3)],

$$\overline{Fh} = f'r, \text{ i.e., } f' = \frac{\overline{Fh}}{r};$$

Hence finally from (75)

$$n = \frac{\overline{Fh}l}{2\pi r^3 \phi t}. \tag{76}$$

This equation, then, expresses n, the coefficient of rigidity, in terms of quantities all of which are easily measurable.

If, as is usually the case, it is found more convenient to twist a *solid* cylinder rather than a hollow one, formula (76) can be transformed to fit the case as follows: Imagine the solid cylinder to be made up of a large number of concentric hollow cylinders of equal thickness and of radii $r_1, r_2, r_3 \ldots r_n$ (see Fig. 45). The moment of force, say $\overline{Fh}_k$, which it is necessary to apply in order to twist any particular hollow cylinder of radius r_k through the angle ϕ, is found from (76), viz.,

Measurement of n from solid cylinder.

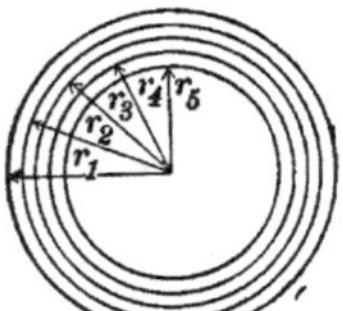

FIGURE 45

$$\overline{Fh}_k = \frac{2\pi n r_k^3 \phi t}{l}. \tag{77}$$

The *total* moment of force $\overline{Fh}$ which must be applied in order to twist the solid cylinder through the angle ϕ is evidently the sum of all the moments $\overline{Fh}_k$. Thus $\overline{Fh} = \overline{Fh}_1 + \overline{Fh}_2 + \overline{Fh}_3 + \ldots \overline{Fh}_n = \frac{2\pi n\phi t}{l}(r_1^3 + r_2^3 + r_3^3 + \ldots r_n^3) = \frac{2\pi n\phi}{l} \times \frac{R^4}{4}$ in which R is the radius of the solid cylinder.* Hence, finally, for a solid cylinder,

$$n = \frac{2\,\overline{Fh}l}{\pi R^4 \phi}, \tag{78}$$

an equation which shows that the twist ϕ produced by a given moment of force, or "torque," $\overline{Fh}$, applied to one end of a cylindrical wire the other end of which is firmly clamped, is directly proportional to the length and inversely proportional to the fourth power of the diameter of the wire.

It is evident from the definition of the coefficient of rigidity n (also called the "*modulus of torsion*") that it is a constant which is characteristic of the *substance* and is independent of the dimensions of the particular wire used. But for a given value of l and R [see (78)], i. e., for each particular wire, the ratio $\frac{\overline{Fh}}{\phi}$ must also be constant (Hooke's Law). This constant of the wire is technically called its "*moment of torsion*," and is represented by the symbol T_o. Thus the equation

Definition of "moment of torsion" T_o.

---

*If the student is not familiar with simple integrations he may take this summation for granted. Elementary integral calculus gives $\int_0^R r^3 dr = \frac{R^4}{4}$, the thickness t in the above expression being the same as dr.

$$T_o = \frac{\overline{Fh}}{\phi} \tag{79}$$

is simply the definition of the *moment of torsion.* Expressed in words, the *moment of torsion of a wire is the constant ratio of the moment of the restoring force exerted by a twisted wire and the angle of twist.* If in (79) ϕ = unity (i. e., 1 radian), then $T_o = \overline{Fh}$. Hence the moment of torsion of a wire is sometimes defined as the moment of force required to twist one end of the wire through one radian.

It is evident from (78) and (79) that the modulus of torsion n may be expressed in terms of the moment of torsion T_o and the dimensions l and R. Thus,

$$n = \frac{2\,T_o l}{\pi R^4}. \tag{80}$$

Experiment

Object. (1) To test Hooke's Law for torsion; (2) to determine the moments of torsion of steel wires of different lengths and diameters; (3) to find n, the coefficient of rigidity of steel.

Description. Three steel wires 1, 2, and 3 (Fig. 46) are provided, of which 1 and 2 have the same lengths l (about 1 m.) but different diameters, 1 and 3 the same diameters (about 2.5 mm.) but different lengths. By means of a set screw at s the wires may be clamped rigidly to the grooved and graduated circular wheel C. Displacements produced by the weights W are read off by means of an index attached to the frame. Ball bearings virtually do away with all friction and render possible a high degree of nicety in the readings.

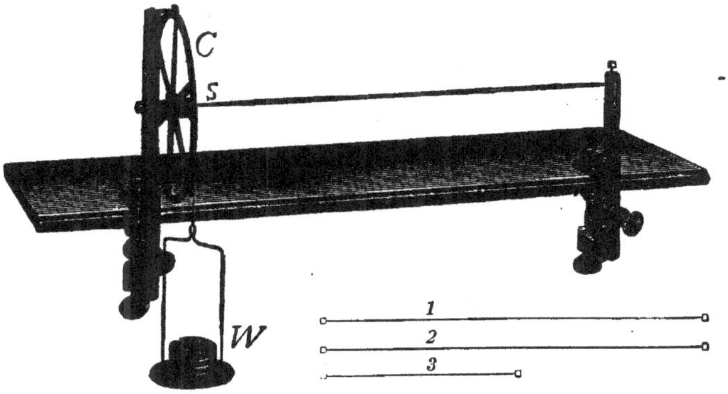

FIGURE 46

Clamp wire 1 in position and take readings first as 100 gm. weights are successively added to the pan, then as they are removed. Repeat with wires 2 and 3. Measure the diameters with the micrometer calipers (see Appendix), taking a mean of a large number of readings at regular intervals along the wire.

Directions.

Use the following method in calculating the mean twist per 100 gm.:

If the total twist due to	6	hundred grams is, say,	4.32
" " " " " "	5	" " " " "	3.61
" " " " " "	4	" " " " "	2.89
" " " " " "	3	" " " " "	2.17
" " " " " "	2	" " " " "	1.44
" " " " " "	1	" " " " "	.73
" " sums =	21	and	15.16

then the most correct value of the twist per 100 gm. which can be obtained from this set of readings is 15.16 ÷ 21 = .722.

This method gives to each observation precisely the amount of consideration which it deserves. Thus it gives a weight of 6 to the observed displacement for 600 gm., a weight of 2 to the observed displacement for 200 gm., etc.

Record

Wire No. 1			No. 2		No. 3		Diameters		
Wts.	Reads.	Difs.	Reads.	Difs.	Reads.	Difs.	No. 1	No. 2	No. 3
0									
100									
200									
300									
400									
500									
600									
500									
400									
300									
200									
100									
0						Means			

$\frac{l \text{ of No. 1}}{l \text{ of No. 3}} =$ —— * $\frac{\text{Mean twist of No. 1}}{\text{Mean twist of No. 3}} =$ —— % of error = ——

$\frac{\text{Diam. of No. 1}}{\text{Diam. of No. 2}} =$ —— $\left(\frac{\text{Mean twist of No. 2}}{\text{Mean twist of No. 1}}\right)^{\frac{1}{4}} =$ —— % of error = ——

Radius of wheel C —— ∴ T_0 for 1 —— for 2 —— for 3 ——

∴ n from 1 —— from 2 —— from 3 —— Mean —— %s difs. ——

Problems

1. If one of the wires had a real diameter of 2.513 mm., but was measured as 2.501 mm., what per cent of error was thus introduced into n?

2. Decide from a study of your observations which one of the quantities involved in n introduces the largest error into the result. Why is it needless to take great pains in measuring the lengths?

3. Show from equation (78) that n would be correctly defined as the moment of force required to twist a cylinder of 1 cm. length and 1 sq. cm. cross-section through 360°.

4. The *moment of torsion* of a particular wire is 7.21×10^6 absolute units; its diameter is 2.732 mm.; its length is 50.1 cm. Find the moment of force required to twist a wire of the same material of 1 mm. diameter and 4 cm. length through 90°.

5. A man grips upon the circumference of a bar 100 in. long and 1 in. in diameter and twists it through 1°. He applies the same force (not the same moment of force) to the circumference of a bar 2 in. in diameter and 80 in. long. Find the twist.

---

* Blanks are for *results* of division.

X

MOMENT OF INERTIA

Theory

It was experimentally shown in Ex. IV that the condition of rotational equilibrium of a rigid body acted upon by the two forces F and W (see Fig. 16) is $Fl = Wl'$. But equilibrium is reached only when the two rates of rotation due to the forces F and W are equal and opposite. It follows, then, from Ex. IV, that the rate at which a force can impart angular velocity to a rigid body is proportional to the product of the force and its lever arm, i. e., to the applied moment of force. Thus, while *linear acceleration* is proportional to the *acting force*, *angular acceleration* is proportional to the acting *moment of force*. Hence "moment of force" bears precisely the same relation to rotary motion which force bears to linear motion.

Motion of rotation.

The inertia of a body is that property by virtue of which it offers resistance to acceleration. The measure of inertia is the resistance offered to unit acceleration; or, since this resistance is always equal to the force producing the acceleration (see scholium), the measure of the inertia of a body is the force necessary to impart to it unit acceleration. This was experimentally proved in Ex. II to be proportional to mass; in the absolute system of units *equal* to mass. Hence a gram of mass has one unit of inertia, two grams of mass two units of inertia, etc.

Inertia.

Moment of inertia is that property of a rotating body by virtue of which it offers resistance to angular acceleration. It is measured by the "moment of force" necessary to impart to the body unit angular acceleration. Thus, a rotating body has unit moment of inertia if it requires the application of a unit moment of force (1 dyne-centimeter) to increase or decrease its angular velocity at the rate of one radian per second; it has two units of moment of inertia if it requires two dyne-centimeters to impart one radian of acceleration, 10 units if it requires 5 dyne-centimeters to impart an acceleration of $\frac{1}{2}$ radian per sec., etc. Symbolically, if I represent moment of

Moment of inertia.

inertia, $\overline{Fh}$ the acting moment of force, and α the angular acceleration produced,

$$I = \frac{\overline{Fh}}{\alpha}. \tag{81}$$

This equation is to be regarded merely as the *definition* of I. It is thus seen that "moment of inertia" is a perfectly definite physical quantity which can be determined for any body whatever by merely applying a known moment of force and measuring the angular acceleration produced. The definition of I might be put in the following form: *Moment of inertia is that physical property in which two rotating bodies agree when it requires the same moment of force to give to each a given angular acceleration.*

Calculation of moment of inertia.

It is evident that I is not proportional to mass alone, as is inertia, for everyday experience teaches that two rotating bodies may have precisely the same mass and yet offer widely different amounts of resistance to the operation of starting or stopping; e.g., two wheels, one of which has its mass concentrated near the axle, the other on the circumference. Thus I is a function both of mass and of the *distribution* of mass, i.e., of the distances of the elements of mass from the axis. In order to calculate I, the moment of force necessary to impart a radian of acceleration must be found in terms of the masses of the particles and their distances from the axis. Take a single particle m_1 at a distance r_1 [$=om_1$] (Fig. 47) from the axis and think of it as moving independently of all the other particles under the action of a force f_1 which gives it a linear acceleration a_1. The second law gives $f_1 = m_1a_1$. The moment of the force f_1 about the axis is f_1r_1.

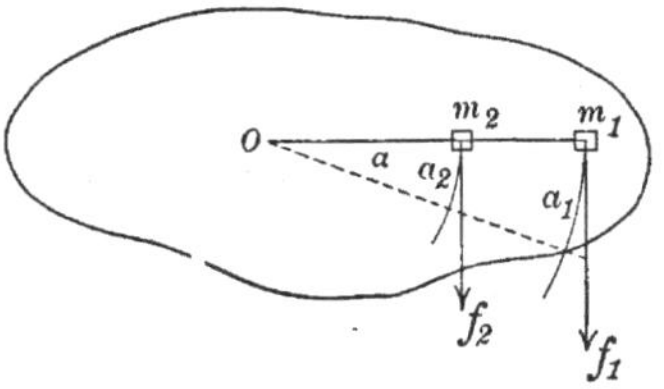

FIGURE 47

Hence $$f_1r_1 = m_1a_1r_1.$$

But since $\frac{a_1}{r_1} = \alpha$ it follows that

$$f_1r_1 = m_1\alpha r_1^2.$$

Similarly the moment of the force f_2 which is necessary to give to m_2 the angular acceleration α about the axis is

$$f_2r_2 = m_2\alpha r_2^2.$$

The total moment of force $\overline{Fh}$ which must be applied to give all the particles of the body the angular acceleration a is manifestly the sum of the moments applied to the several particles. Thus:

$$\overline{Fh} = f_1r_1 + f_2r_2 + \text{etc.} = a\,(m_1r_1^2 + m_2r_2^2 + \text{etc.}) = a\Sigma mr^2 \qquad (82)$$

or

$$\frac{\overline{Fh}}{a} = \Sigma mr^2, \qquad (83)$$

i. e., the moment of force necessary to impart unit angular acceleration is Σmr^2. But this is by definition I (see 81).

Hence

$$I = \Sigma mr^2. \qquad (84)$$

In order, then, to calculate I for any body, it is necessary to multiply the mass of each particle in the body by the square of its distance from the axis of rotation, and then to find the sum of all these products. For an irregular, non-homogeneous body, this would evidently be an impossible undertaking. Hence for such bodies the moment of inertia can not be calculated. It can only be obtained by direct experiment, i. e., by applying a known moment of force and observing the angular acceleration produced (or by means of some experiment which is equivalent to this). But for certain regular, homogeneous bodies, it is possible to perform the summation indicated and hence to check an experimental value of I by means of a calculated one. This summation is done with the aid of the integral calculus. Only a few results of such summation will be indicated here.

The moment of inertia of a uniform cylinder of radius R and mass M rotating about its own axis is

$$I = \frac{MR^2}{2}.\text{*} \qquad (85)$$

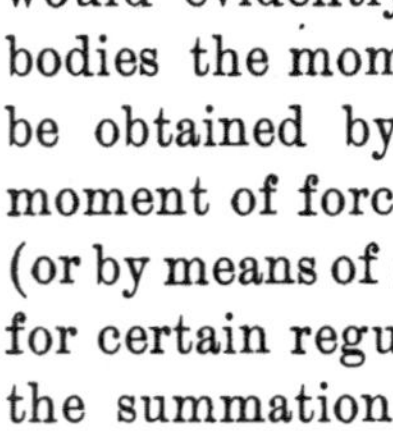

FIGURE 48

* This may be obtained as follows: If σ represent the density of the cylinder and l its length, then with the use of polar coördinates (see Fig. 48), an element m of mass may be taken as

$$m = \sigma lr\,dr\,d\theta$$

Hence

$$I = \Sigma mr^2 = \Sigma\sigma l\,d\theta\,r^3dr = \sigma l\int_0^{2\pi}\int_0^R r^3dr\,d\theta = \frac{\sigma l}{2}\pi R^4 = \frac{MR^2}{2}.$$

The other integrations are somewhat more complicated and will not here be attempted.

The moment of inertia of a uniform sphere of radius R and mass M rotating about an axis passing through its center is

$$I = \frac{2}{5} MR^2. \tag{86}$$

The moment of inertia of a uniform rectangular bar of length l, width a, thickness b, and mass M, rotating about an axis parallel to b, and passing through its center of gravity is

$$I = \frac{M}{12}(l^2 + a^2). \tag{87}$$

Comparison of linear and rotary motions.

From a comparison of the equations $f = ma$ and $\overline{Fh} = I\alpha$ it appears that I bears precisely the same relation to rotary motion which mass bears to linear motion. I is therefore sometimes called the mass of rotary motion. Whenever m appears in an expression which relates to linear motion I always appears in the corresponding expression which relates to rotary motion. Thus, for example, the kinetic energy of linear motion is $\frac{1}{2}mv^2$. The kinetic energy of rotary motion is $\frac{1}{2}I\omega^2$, ω being the angular velocity of the rotating body. This may be shown as follows: The kinetic energy ke_1 of the particle m_1 (see Fig. 47) is $\frac{1}{2}m_1v_1^2$.

But $$v_1 = \omega r_1.$$

Hence $$ke_1 = \frac{1}{2}m_1r_1^2\omega^2.$$

Now the total kinetic energy KE of the body must be the sum of the kinetic energies of its parts.

Therefore $$KE = \Sigma\frac{1}{2}mr^2\omega^2 = \frac{1}{2}I\omega^2. \tag{88}$$

Q. E. D.

Experiment.

Object.

To determine the moment of inertia of a circular disk by applying a given moment of force and measuring the corresponding angular acceleration, and to compare the result with the theoretical value of I.

Description.

A disk weighing several kilograms is mounted upon ball bearings so as to rotate with very little friction about its own axis (see Fig. 49). While a weight m imparts to the disk an angular acceleration α, an electrically driven fork of known period writes a trace upon the blackened face of the disk. The angular acceleration α is determined precisely as were the linear accelerations in Exs. I and II, i. e., by subtracting successive angular distances traversed during successive equal intervals of time. These angular intervals are obtained in degrees from readings made upon a circular scale with which the unblackened face of the disk is provided. By means of the rack and pinion *s*, the frame which carries the fork may be moved through ways in a direction parallel to the face of the disk, so that the traces made during successive revolutions of the disk need not interfere. The screw *s′* shifts the whole disk in the direction of its axis, and thus makes it easy to secure a suitable pressure of the stylus against the blackened face of the disk.

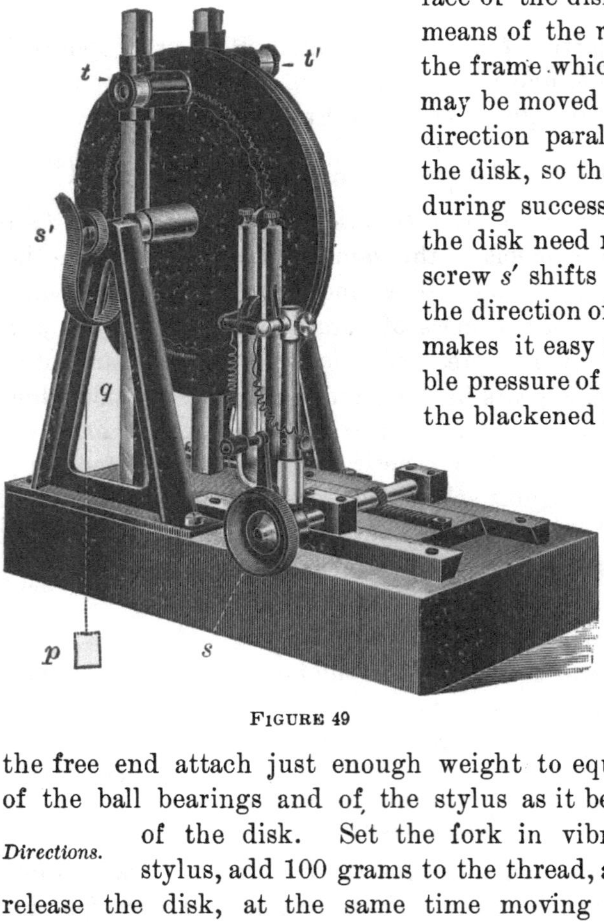

FIGURE 49

Wrap a fine thread three or four times around the circumference, attaching one end to the disk by means of a small bit of wax. To the free end attach just enough weight to equalize the friction of the ball bearings and of the stylus as it bears upon the face of the disk. Set the fork in vibration, adjust the stylus, add 100 grams to the thread, and then suddenly release the disk, at the same time moving forward the fork by means of *s*. As soon as *m* touches the floor, stop the disk, remove it from the frame, and carefully mark off the trace as in Ex. I, taking a group of 50 waves as the unit. Replace the disk in the frame, set the cross-hairs of the low-power microscope *t* upon the limiting mark of the first group of waves, and take the reading of the cross-hairs of the microscope *t′* upon the circular scale graduated upon the unblackened face of the disk. Then turn the disk until the second mark comes underneath the cross-hairs and read again. From such readings α is easily obtained.

Directions.

It must, of course, be expressed in radians (see definition of moment of inertia). Take at least two traces, using different masses for m; e. g., let $m_1 = 100$ gm., $m_2 = 200$ gm.

Calculation of $\overline{Fh}$.

The force F which produces the rotation of the disk is evidently the tension in the thread q. This is not the force acting upon the mass m, viz., mg, for a part of the acting force mg is expended in producing the acceleration a which is imparted to m. By the scholium to the Third Law, $mg = F + ma$, or $F = m(g - a)$. Since the acceleration of the weight is the same as the acceleration at the circumference of the disk, $a = \alpha R$, R being the radius of the disk and α its angular acceleration in radians. Hence the acting moment of force $\overline{Fh}$ may be found from the measurement of the three quantities α, R, and m. $\overline{Fh}$ must of course be expressed in dyne-centimeters. From $\overline{Fh}$ and α, I is at once obtained (see definition of I). The theoretical value of I, see (85), involves only the mass M and the radius R of the disk. M is to be obtained by weighing upon the platform scales.

Record

1ST DETERMINATION			2D DETERMINATION		
m_1 ———	Chosen interval ——— vibs.		m_2 ———	Fork rate ———	Interval ———
Angle Read'gs	Angular Spaces	Angular Acc's	Angle Read'gs	Angular Spaces	Angular Acc's
———			———		
———	———	———	———	———	———
———	———	———	———	———	———
———	———	———	———	———	———
———	———	———	———	———	———
———	———	———	———	———	———
———	———	———	———	———	———
———	———		———	———	
	Mean acc. = ———			Mean acc. = ———	
$R =$ ———	$\therefore I =$ ———		$\therefore I =$ ———	Mean $I =$ ———	
$M =$ ———	$\therefore$ calc'd $I =$ ———			% error in obs'd $I =$ ———	

Problems

1. A constant pull of 200 kgm. acting on the circumference of a wheel of 1 m. radius imparts in 30 sec. a speed of two revolutions per second. Find I for the wheel.

2. Find what part of the kinetic energy of a rolling solid cylinder is energy of translation, and what part energy of rotation.

The latter energy, viz., $\frac{1}{2}I\omega^2$, can in this case be expressed in terms of the mass M of the cylinder and its velocity of translation v. (See 85.) A simple relation exists between v and ω.

3. Solve Problem **2** for a rolling hoop; for a rolling solid sphere.

4. Find what relation exists between the velocities acquired by a solid cylinder in sliding without friction down an inclined plane and in rolling without slipping down the plane.

Equate in each case potential energy at top to total kinetic energy at bottom.

5. The wheel used in the falling body machine (Ex. II) has a radius of 5 cm. and moment of inertia of 275 gm. cm.^{2} * Find what mass at the circumference would offer the same resistance to an accelerating force and therefore what number of grams should be added to the mass of the frame in order to allow for the presence of the wheel.

6. Find as in Problem **4** the relation between the velocities of solid spheres sliding and rolling down an inclined plane.

See equation (86) and the suggestion under Problem **4**.

7. A bullet weighing 5 gm. and moving with a velocity of 100 meters per second in the direction *ab* (see Fig. 50), strikes the projection *b* of a fixed wheel whose moment of inertia is 200,000 gm. cm.^{2},* and whose radius is 20 cm. Find the number of revolutions per second communicated to the wheel.

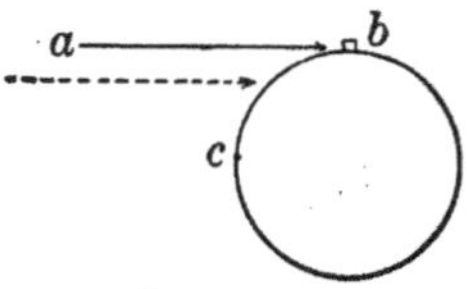

FIGURE 50

---

* In order to understand the meaning of this symbol it will be necessary to consider briefly the origin and use of dimensional formulae. The dimensional formula for any quantity is simply the symbol which represents the way in which the *fundamental units*, i.e., the units of Mass, Length and Time, enter into the definition of that quantity; e. g., note the following dimensional formulae:

Since velocity is length divided by time, $v = \frac{L}{T} = LT^{-1}$.

Since acceleration is velocity divided by time, $a = \frac{L}{T^2} = LT^{-2}$.

Assume inelastic impact. Then if v represent the initial velocity of the bullet, ω the angular velocity of bullet and wheel after impact, and t the duration of the impact, the velocity lost by the bullet is evidently $v - \omega R$, R being the radius of the wheel. Hence the mean force acting between bullet and wheel during the impact is $f\ [= ma] = \frac{m(v - \omega R)}{t}$. The moment of this force imparts to the wheel a mean angular acceleration α such that $\alpha t = \omega$. From these equations and that which defines I, ω may be found.

8. Find the number of revolutions per second which would have been imparted to the wheel if the bullet had moved along the dotted line (see Fig. 50) and struck the wheel at a point midway between b and c.

9. A hoop and a solid disk of the same diameter start down a hill together. Which will reach the bottom first? Find the ratio of their velocities at the bottom.

Assume in each case rolling without slipping and neglect air resistance. See Problem 4.

10. Why can a heavy man on a bicycle always coast faster than a light man on the same wheel?

In answer disregard friction.

Since force is mass multiplied by acceleration, $f = \frac{ML}{T^2} = MLT^{-2}$.

Since moment of force is force multiplied by length, $\overline{Fh} = \frac{ML^2}{T^2} = ML^2T^{-2}$.

Since angular accel. is linear accel. divided by length, $\alpha = \frac{1}{T^2} = T^{-2}$.

Since moment of inertia is moment of force divided by angular acceleration $I = ML^2$.

This last result might have been seen at once from the fact that it has been shown that $I = \Sigma mr^2$. Enough has been said to show the method of procedure for the derivation of dimensional formulae and the meaning of such formulae.

Now when units have been defined but have been given no particular names it is customary to write the dimensional formula in place of a name, replacing in this formula the general symbols M, L, T, by the particular units of mass, length, and time which have been used in the definition of the unit under consideration. Thus if it were desired to say that the moment of inertia of a body was 275 units and to explain that in the definition of these units the gram and the centimeter were taken as the fundamental units of mass and length and that these were involved in such a way in the definition that the dimensional formula for moment of inertia was $I = ML^2$ it would only be necessary to write "moment of inertia of body $= 275$ gm. cm.^{2}"

11. Will increasing the weight of the tires increase or decrease the coasting speed of a bicycle? What effect will increasing the weight of the frame have upon the coasting speed?

12. A clay ball weighing 50 gm., and moving with a velocity of 30 meters per second, struck and stuck to the end of a rectangular bar 1 meter long and 5 cm. square which was pivoted at its center of gravity *o* (see Fig. 51). If the weight of the bar were 5 kilos and the motion of the ball were at right angles to the length of the bar, what number of revolutions per second would be communicated to the bar?

FIGURE 51

See equation (87).

XI

SIMPLE HARMONIC MOTION

Theory

Definition of S.H.M.

The two very simple forms of motion thus far considered, viz., uniform and uniformly accelerated, belong to the general class of *non-periodic* motions.

Among periodic motions the simplest and most important type is so-called *simple harmonic motion.* This is defined as motion in which the oscillating body is at every instant urged toward some natural position of rest with a force which varies directly as its distance from that position. Thus suppose a particle to be moving back and forth over the path ab (see Fig. 52), under the action of a force which has its origin in o, and let the law of action of the force upon the particle be expressed by the equation,

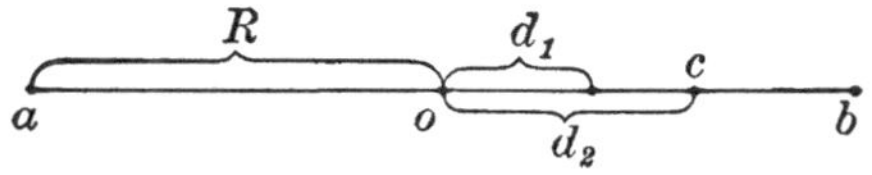

FIGURE 52

$$\frac{f_1}{f_2}\left(=\frac{ma_1}{ma_2}\right)=\frac{d_1}{d_2}, \tag{89}$$

in which f_1 and f_2 represent the forces acting upon the particle when it is at the distances d_1 and d_2 respectively from o, and a_1 and a_2 represent the accelerations toward o of the particle when it is at these distances. The last equation may be written in the form,

$$\frac{a_1}{d_1}=\frac{a_2}{d_2}=\frac{a_3}{d_3} \text{ etc.};$$

or, in general, if this constant ratio of the acceleration to the displacement be denoted by the symbol k, the equation which defines simple harmonic motion [see (89)] may be written in the form,

$$a = kd. \tag{90}$$

Since $f = ma$, (90) may evidently be written in the form,

$$f = mkd. \tag{91}$$

Any one of equations (89), (90), or (91) may be taken as the definition of simple harmonic motion. These equations express simply a proportionality between force and displacement. But, in view of a very simple relation which exists between the characteristic constant k of the motion and the *period of vibration* of the system, it is possible to express the characteristic equation of simple harmonic motion in still another form. The evaluation of k in terms of the period will be made in three steps, as follows:

k in terms of v, R and d.

1. The first step will consist in finding k in terms of the half length R of the path of the particle, the velocity v which it has at any particular point of this path, for example at c, and the distance d of the chosen point c from o. Since the particle comes to rest at the end of its path, i. e., at b, the velocity v is acquired while it is moving from b to c under the action of the force which has its origin in o. Hence, by equation (30), p. 43, the kinetic energy which the particle has acquired when it reaches c, viz., $\frac{1}{2}mv^2$, is equal to the work done by the force in moving it from b to c. Now it may be seen from (91) that the value of the force acting upon the particle when it is at b is mkR, while its value at c is mkd. Since, by the definition of simple harmonic motion, the force is always proportional to the distance of the particle from o, the *mean* value of the force acting upon it while it is moving from b to c is $\frac{mkR + mkd}{2}$. The distance bc is equal to $R - d$.

Hence $$\tfrac{1}{2}mv^2 = \frac{mkR + mkd}{2} \times (R - d);$$

or $$v^2 = k\,(R^2 - d^2). \qquad (92)$$

k in terms of R and S.

2. The second step will consist in expressing k in terms of R and the speed S with which a shadow cast by the particle moves about the circumference of a circle drawn upon ab as a diameter. Imagine that a distant source of light, situated directly below ab (Fig. 53), casts a shadow of the moving particle upon the circumference ahb. If the velocity v of the particle at any point

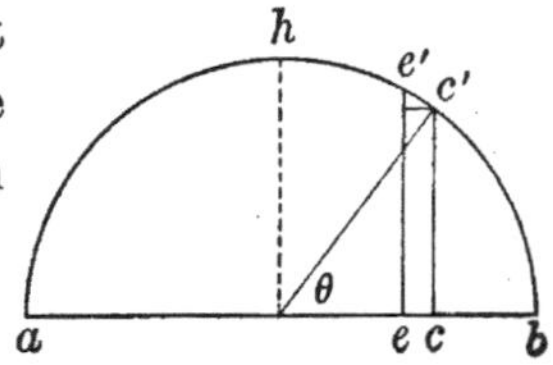

FIGURE 53

c be represented by the line ce, then the velocity S of the shadow upon the circumference is represented by $c'e'$. But

$$\frac{ce}{c'e'} = \sin\theta.$$

Hence
$$v = S\sin\theta.$$

Or
$$v^2 = S^2\sin^2\theta = S^2(1-\cos^2\theta) = S^2\left(1-\frac{d^2}{R^2}\right).$$

Or
$$v^2 = \frac{S^2}{R^2}(R^2-d^2). \qquad (93)$$

From (92) and (93) there results at once

$$k = \frac{S^2}{R^2} \qquad (94)$$

This equation shows that the shadow travels with *uniform speed* about the circumference, for the expression for S is independent of d. It involves only the two constants k and R.

3. Since the shadow has a constant speed, the time which it requires to move over the semicircumference bha is the length of

k in terms of the half-period t.

this path divided by the speed, viz., $\frac{\pi R}{S}$. But this time is manifestly the time t which the particle consumes in moving from b to a, i. e., it is the half-period of the vibration. Thus,

$$t = \pi\frac{R}{S}. \qquad (95)$$

From (94) and (95) it follows at once that

$$k = \frac{\pi^2}{t^2}. \qquad (96)$$

Characteristic equations of S.H.M.

Hence the characteristic equation (90) of simple harmonic motion becomes

$$a = \frac{\pi^2}{t^2}d, \qquad (97)$$

and (91), which simply represents the combination with (90) of the equation $f = ma$, becomes

$$f = m\frac{\pi^2}{t^2}d, \quad \text{or} \quad t = \pi\sqrt{\frac{m}{\frac{f}{d}}}. \qquad (98)$$

$\frac{f}{d}$ is called the *force constant* of the system considered.

If it is known, t can at once be determined. Thus, if the oscillatory motion is due to the elasticity of a spring, the force constant $\frac{f}{d}$ can be determined once for all by observing the force required to stretch the spring through a given distance. Then the period of oscillation of the system when any mass m is hung from the spring can be calculated from (98).

The force constant.

The enormous importance of simple harmonic motion in the study of Physics arises in part from the fact that all vibrations arising from the elasticity of matter are cases of simple harmonic motion. For, by Exs. VIII and IX, the restoring forces called into play by any sort of strains in material bodies are proportional to the displacements, so long as the limits of perfect elasticity are not exceeded. Thus the vibrations of weights suspended from springs, the vibrations of tuning forks, of the strings of any stringed instrument, of masses vibrating under the torsion of a wire, are all cases of simple harmonic motion.

S.H.M. in nature.

A body oscillating about an axis has *simple harmonic motion of rotation* when each of its particles moves with *linear* simple harmonic motion. Thus let the line om (Fig. 54) oscillate between the positions oc and ob, and let each point n of the line follow the equation of linear simple harmonic motion, viz., $a = \frac{\pi^2}{t^2}d$. Since the linear quantities d [$= bc$] and a [$= mp$] are related to the corresponding angular quantities θ and α by the equations $\frac{d}{R} = \theta$ and $\frac{a}{R} = \alpha$ [$R = om$], the equation $a = \frac{\pi^2}{t^2}d$ may be at once transformed into

Characteristic equations of S.H.M. of rotation.

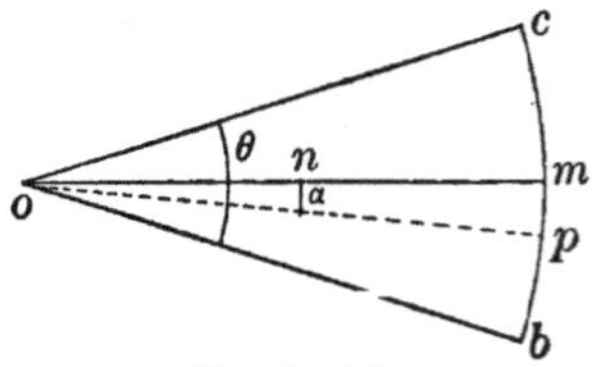

FIGURE 54

$$\alpha = \frac{\pi^2}{t^2}\theta. \tag{99}$$

The combination of this simple harmonic rotational equation with the general rotational relation $\overline{Fh} = I\alpha$ [see (81) p. 79], gives

$$t = \pi\sqrt{\frac{I}{\frac{\overline{Fh}}{\theta}}}. \tag{100}$$

$\frac{\overline{Fh}}{\theta}$ is the *force constant* of the rotational system. In the case in which the rotation was due to torsion this constant ratio $\frac{\overline{Fh}}{\theta}$ was given the name "moment of torsion" and was represented by the symbol T_o (see Ex. IX).

Force constant of S.H.M. of rotation.

It is evident from (100) that the moment of inertia I of any body may be experimentally determined by suspending it from a wire of known moment of torsion T_o and observing the half-period of vibration t_1. Thus,

I and T_o from torsional vibrations.

$$t_1 = \pi\sqrt{\frac{I}{T_o}}, \qquad \text{or} \qquad I = \frac{T_o t_1^2}{\pi^2}. \tag{101}$$

Or again, if the moment of torsion T_o of the suspending wire be not known, the observation first of t_1 and then of a second period t_2 in which the system is caused to vibrate by the addition to I of a *known* moment of inertia I_o, furnishes all necessary data for determining either I or T_o. Thus the half-period of the system after the addition of I_o is given by

$$t_2 = \pi\sqrt{\frac{I + I_o}{T_o}}. \tag{102}$$

The elimination of T_o from (101) and (102) gives

$$I = \frac{I_o t_1^2}{t_2^2 - t_1^2}. \tag{103}$$

Or the elimination of I from (101) and (102) gives

$$T_o = \frac{I_o \pi^2}{t_2^2 - t_1^2}. \tag{104}$$

Experiment

1. To determine the force constant of a spiral spring, and to compare the observed and calculated values of the periods of the spring for different loads. 2. To calculate the periods of several "torsion pendulums" from the known T_o's of the suspending wires, and to compare with observations. 3. To determine I and T_o by the addition to a vibrating system of a known moment of inertia I_o.

Object.

DIRECTIONS.—1. First test Hooke's Law for the spring and determine its force constant by observing the elongations produced by the successive addition of 100 gm., 200 gm., 300 gm., 400 gm., beginning with a load of 50 gm. A graduated mirror placed behind the spring (see Fig. 55) enables the position of the index for any value of m to be accurately determined. In taking a reading place the eye so that the image of the tip of the index is brought into line with the tip of the index itself. In computing the mean elongation per hundred grams use the method outlined on page 76. In computing the force constant $\frac{f}{d}$ express all forces in dynes, all lengths in centimeters.

Force constant of spiral spring.

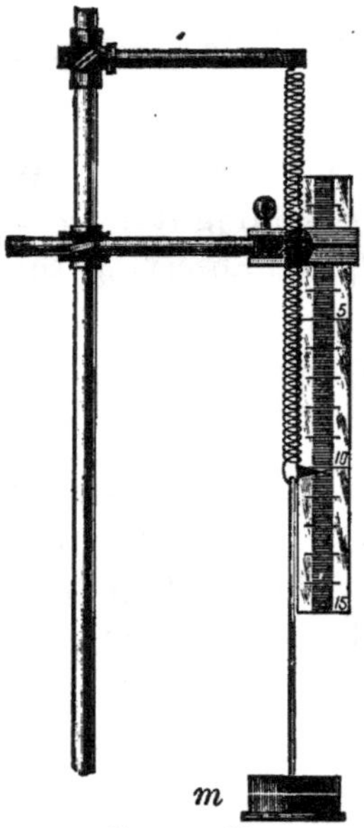

FIGURE 55

Now replace m by other masses, e. g., 150 gm., 250 gm., 350 gm., and, from the force constant just determined, calculate in each case what should be the period of the suspended system when it is set into vertical oscillation. Compare these theoretical values of the periods with observed values obtained by taking with a stop-watch the time of 50 vibrations.

Periods for different loads.

2. The torsion pendulum here used consists of a large disk suspended as in Fig. 56 from one of the wires whose force constant (moment of torsion) was found in Ex. IX. First find the weight of the disk by means of the platform scales, then measure its diameter and compute its moment of inertia I [see (85) p. 80]. From I and the value of T_0 found in Ex. IX compute the period. Compare this with the observed period obtained by taking with a stop-watch the time of 25 vibrations. In setting the disk into oscillation, do not twist the wire through an arc *of more than 5° or 10°*.

Periods of torsion pendulums.

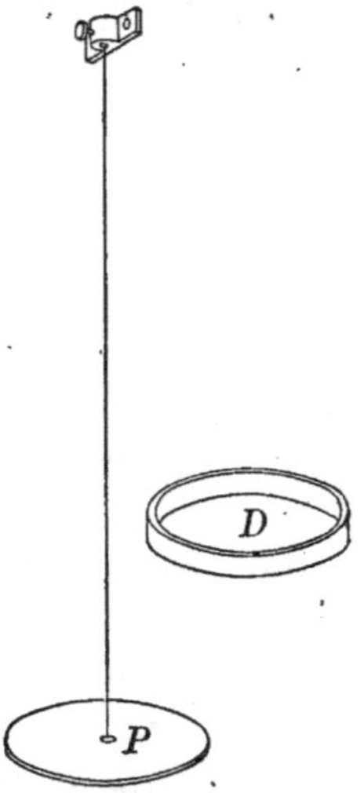

FIGURE 56

3. The determination of T_0 by the addition of a known moment of inertia I_0 is accomplished by adding to the plate P (Fig. 56) the large brass ring D, and observing the new period t_2. T_o is then given by (104). I_o can of course be easily calculated from its mass and mean diameter [see (84) p. 80]. Care must be taken to make the wire pass through the center of the ring; otherwise the calculated value of I_o will be incorrect. Determine in this manner T_o for each of the three wires used in Ex. IX, and then calculate from each wire the coefficient of rigidity n of steel. [See (80) p. 75].

T_o and I_o from added moment of inertia.

From the same observations calculate with the aid of (103) the moment of inertia of the brass disk, and compare with the theoretical value (see 85).

Record

1.

Added W'ts	Scale Read's	Dif's
0	———	
		———
100	———	
		———
200	———	
		———
300	———	
		———
400	———	

∴ mean d per 100 gm. = ———

∴ force cons't of spring = ———

Masses	t (obs.)	t (calc.)	% error
150	———	———	———
250	———	———	———
350	———	———	———

2. Mass of disk P = ——— Radius = ——— ∴ I = ———

Wire 1, T_o (from IX) ——— ∴ t_1 calc. ——— t_1 obs. ——— % error ———

Wire 2, T_o (from IX) ——— ∴ t_1 calc. ——— t_1 obs. ——— % error ———

Wire 3, T_o (from IX) ——— ∴ t_1 calc. ——— t_1 obs. ——— % error ———

3. Mass of ring D = ——— Mean radius = ——— ∴ I_o = ———

Wire 1, t_2 = ——— ∴ T_o = ——— ∴ n = ——— % dif. from mean = ———

Wire 2, t_2 = ——— ∴ T_o = ——— ∴ n = ——— " " " " = ———

Wire 3, t_2 = ——— ∴ T_o = ——— ∴ n = ——— " " " " = ———

I from 1 ——— from 2 ——— from 3 ——— Mean ——— % error ———

Problems

1. Within a solid sphere of uniform density the force varies directly as the distance from the center. If the earth were such a sphere, and if a hole passed completely through it along a diameter, how long a time would be required for a body dropped through the hole to reach the other side?

Take the radius of the earth as 4000 miles.

2. A horizontal wire one meter long clamped at both ends is set into vibration in a vertical plane. The amplitude at the mid-

dle is 4 mm. Find the shortest period which is permissible if the rider at the mid-point is at no instant to lose contact with the wire.

3. Show that the apparent motion of a bright point on the rim of a distant wheel, rotating at uniform speed about an axis at right angles to the observer's line of sight, is simple harmonic.

XII

DETERMINATION OF "g"

Theory

Let Fig. 57 represent any irregular body which is oscillating under the action of the force of gravity about a horizontal axis at o. Let l be the distance from o to c the center of gravity of the body.

Moment of force restoring a pendulum.

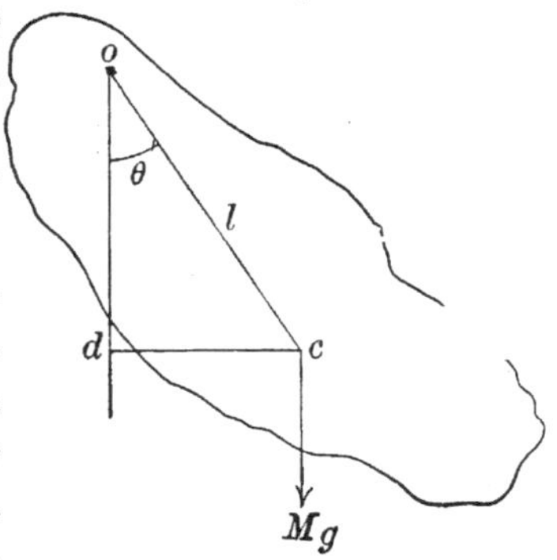

FIGURE 57

In order to find the law which governs the motion of the body it is necessary to express the moment of force $\overline{Fh}$ which is acting upon the body at any instant in terms of the angular displacement at that instant from the position of rest. If M be the mass of the body, the total force acting upon it is Mg dynes, and this force is applied at the center of gravity c (see Ex. IV, pp. 34, 35). The moment of this force is therefore $Mg \times dc$. But since $dc = l \sin \theta$, it follows that

$$\overline{Fh} = Mgl \sin \theta. \tag{105}$$

This equation shows that the motion is *not* simple harmonic, for the restoring moment $\overline{Fh}$ is *not* proportional to the displacement θ, but to $\sin \theta$. Nevertheless, as θ approaches zero, $\sin \theta$ approaches θ. Hence in the limit, i. e., in the case of vibrations of infinitely small amplitude, pendular motions follow the law of simple harmonic motion. The simple harmonic formula may then be applied to pendulum problems if only the arc be kept so small that the error introduced by the approximation $\sin \theta = \theta$ is smaller than the necessary observational errors of the experiment. This means that in the following experiment θ should not exceed 5°. Under these conditions,

The period of a compound pendulum.

then, the pendulum formula is

$$\overline{Fh} = Mgl\,\theta;$$

i. e., the force constant of the motion, viz. $\frac{\overline{Fh}}{\theta}$, is equal to Mgl. Substitution in the S. H. M. formula (100) gives, then,

$$t = \pi \sqrt{\frac{I}{Mgl}}. \tag{106}$$

This is the general formula for the compound pendulum. If the pendulum be merely a particle suspended from a weightless thread of length l, then

Period of simple pendulum.

$$I\ (= \Sigma mr^2) = Ml^2. \tag{107}$$

Therefore, for such a pendulum, (106) becomes

$$t = \pi \sqrt{\frac{l}{g}}. \tag{108}$$

Important definitions.

The *length* of a compound pendulum is defined as the length of the simple pendulum which has the same period.

The *center of oscillation* of a compound pendulum is the position of the particle which is oscillating naturally, i. e., just as it would oscillate if it alone were suspended as a simple pendulum from the point of support.

The *radius of gyration* k of any rotating body is the distance from the axis at which the whole mass M might be concentrated without changing the value of the moment of inertia I. Thus the equation,

$$I = \Sigma mr^2 = Mk^2 \tag{109}$$

defines the radius of gyration k.

Experiment

Object.

To find g by determining the length and the period of a simple pendulum.

A simple pendulum is chosen for the determination, because the length of such a pendulum can be measured directly, while the length of a compound pendulum is not easily obtainable. The time measurement consists in comparing, by the *method of coincidences*, the period of the unknown simple pendulum with that of a compound pendulum of known period. The electric circuit of the battery B (see Fig.

Method in theory.

58) is completed through the electro-magnet E, the contact points c and d, and the two pendulums A and C. The length of the simple pendulum A is adjusted until its period is nearly the same as that of the known pendulum C. When both pendulums strike the mercury contacts at precisely the same instant, the click of the sounder (or the stroke of the bell) E is heard. Since the pendulums have slightly different periods, no further sound is heard until the faster pendulum has gained one half-vibration upon the slower, when the conditions are right for another click. Theoretically, an observation of the interval elapsing between these two successive coincidences is sufficient for the determination of the half-period t of the unknown pendulum. For, division of the time interval between coincidences by the half-period of C gives the number of vibrations made by C during the interval. The addition (or subtraction) of 1 to this number gives the number of half-periods of A during the same interval. Hence the division of the length of the interval by this number gives the half-period sought, viz. t.

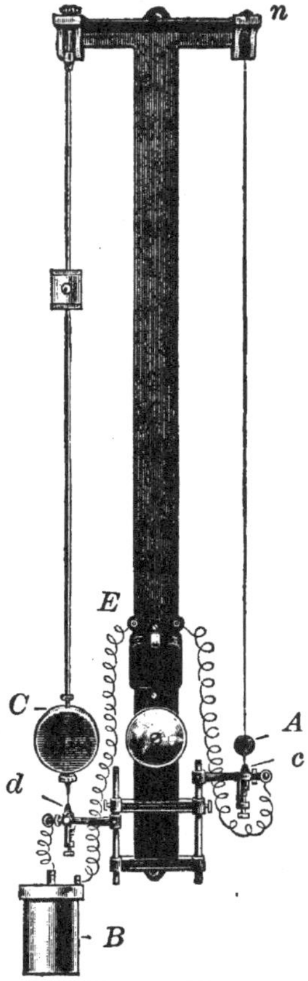

FIGURE 58

However, on account of the difficulty of observing accurately the exact instant of a coincidence, it is preferable to observe the time interval between two coincidences which are a considerable distance apart, e. g., the interval between the 1st and the 30th coincidence. The calculation is then made precisely as outlined, save that the 1 is replaced by 29. This reduces the observational error to $\frac{1}{29}$th its former value. It is not necessary, however, to watch the pendulums through 30 coincidences. For, if the interval between two successive coincidences be somewhat accurately observed, the number of coincidences which have occurred within any longer interval bounded by two coincidences must evidently be the nearest whole number obtained by dividing the long interval by

Method in practice.

the time between two *successive* coincidences. The divisor, however, must be determined with such exactness as to remove all uncertainty as to what that whole number is.

The difficulty of determining the exact instant of a coincidence is increased by the fact that, on account of the finite width of the contact points *c* and *d*, the clicks will often continue through a number of swings. The mean time between the first and last click is then taken as the instant of coincidence.

First very carefully adjust *c* and *d* until, when *A* and *C* are at rest, each pendulum point touches the middle of its contact.

Directions.

Then make battery connections as in the diagram. Next by means of a thread tie back pendulum *A* a distance of about 2 inches and set it into vibration by burning the thread. If a click is heard at every passage, set the heavier pendulum *C* into vibration, giving it an amplitude somewhat smaller than *A*'s. For the determination of *t* use an ordinary watch, but not one which has an error of more than 10 sec. per day. Take, as accurately as possible, the times of the first four coincidences; then allow the pendulum to swing for 20 or 30 minutes and take the time of another coincidence. During the interval watch the pendulums very carefully to see which is the faster, observing just after a coincidence and noting which reaches the end of its swing first. The length of *A* may be taken, without appreciable error, as the distance from the knife-edge *n* to the center of the ball. Leaving the pendulum in place, measure the diameter of the ball with the vernier-calipers (see Appendix), and the distance from the top of the ball to the knife-edge with a meter-stick.

Record

	h.	m.	s.	Coinc. perfect at h.	m.	s.	Interval m.	s.
1st coincidence began at	——	——	——					
" " ended at	——	——	——	——	——	——		
							——	——
2d " began at	——	——	——					
" " ended at	——	——	——	——	——	——		
							——	——
3d " began at	——	——	——					
" " ended at	——	——	——	——	——	——		
							——	——
4th " began at	——	——	——					
" " ended at	——	——	——	——	——	——		
Last " began at	——	——	——					
" " ended at	——	——	——	——	——	——		

Interval bet. 1st and last —— No. coincidences in interval = ——

No. vib's of C in interval = —— No. vib's of A = —— $\therefore t =$ ——

Length of wire —— Diam. of ball —— $\therefore$ length of pend. —— $\therefore g =$ ——

Problems

1. The moment of inertia of a long and thin cylindrical body, oscillating about one end, is $\frac{1}{3}ML^2$, L being the length of the body and M its mass. Find the radius of gyration of the body in terms of L. Find the length of the simple pendulum which has the same period.

See (109) and (106).

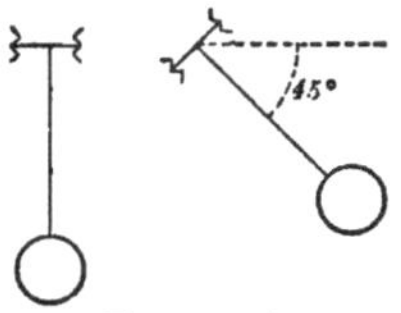

FIGURE 59

2. A rigid pendulum oscillating about a horizontal axis has a period of .75 sec. Find its period when the axis is inclined 45° to the horizontal. (See Fig. 59.)

3. The period of a pendulum at the sea level was 1.002 sec. It was carried to the top of a mountain and the period found to be 1.005 sec. Find the height of the mountain.

XIII

THE LAW OF CENTRIPETAL FORCE

Theory

Acceleration has been defined as rate of change of velocity. But velocity is a *directed* quantity (a so-called vector) and may change either in direction, or in mere numerical value, or in both. The term *speed* has been agreed upon to denote the *numerical value* of the velocity. If, then, (case 1), the velocity of a body at any time is represented by the line or_1 (see Fig. 60), and if its velocity after a lapse of t seconds is represented by or_3, then there has been no change in direction, but only a change in speed, and the mean acceleration has been $\frac{r_1r_3}{t}$. The expression for the acceleration at the instant at which or_1 represents the velocity is evidently the limit of this quantity as t approaches zero; thus,

Acceleration corresponding to change in speed alone.

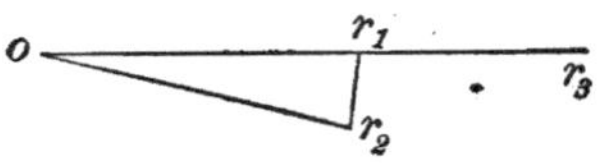

FIGURE 60

$$a = \left(\frac{r_1r_3}{t}\right)_{t \doteq 0}. \tag{110}$$

But if, (case 2), after the lapse of the t seconds, the velocity is represented not by or_3 but by or_2, a line equal in length to or_1, then there has been no change in speed, but only a change in *direction*, and, by the principle of composition of directed quantities (see Ex. III, p. 22), the new velocity or_2 is equivalent to the two simultaneous velocities or_1 and r_1r_2, i. e., to the old velocity or_1 plus a new velocity r_1r_2. r_1r_2 is then the line which represents the gain in velocity during the time t. Hence the mean acceleration during the interval t has been $\frac{r_1r_2}{t}$, and the value of the acceleration at the instant at which or_1 represents the velocity is

Acceleration corresponding to change in direction alone.

$$a = \left(\frac{r_1r_2}{t}\right)_{t \doteq 0}. \tag{111}$$

Since, as t approaches zero, r_1r_2 becomes more and more nearly perpendicular to or_1, it is evident that in the limit represented by (111) the acceleration a is *at right angles* to the velocity or_1. Now in Ex. II *force* was defined as that which changes the motion of a body, and, by the Second Law, whether the change is one of direction or of speed, the measure of the force is always ma, and the direction of the force is the direction of the vector a. Thus, in case 1, the force $ma\left[= m\left(\frac{r_1r_3}{t}\right)_{t \doteq 0}\right]$ acts in the direction of the velocity, i. e. in the direction or_1. In case 2 the force $ma\left[= m\left(\frac{r_1r_2}{t}\right)_{t \doteq 0}\right]$ acts in the direction $(r_1r_2)_{t \doteq 0}$, a direction which is rigorously perpendicular to the velocity or_1.

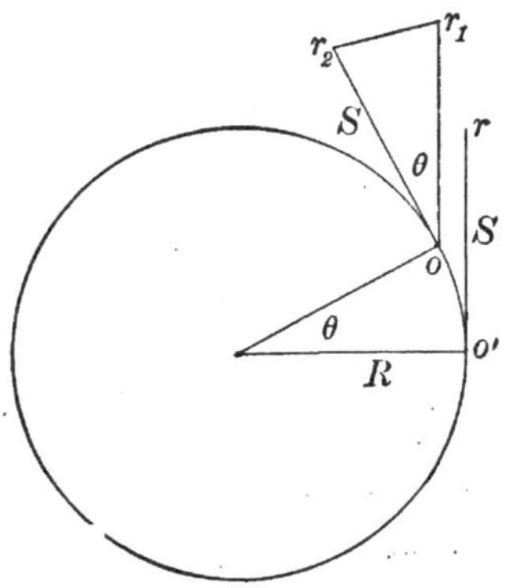

FIGURE 61

Apply these principles to the consideration of the case of a body m moving with uniform speed S upon the circumference of a circle. (See Fig. 61). Let or_2 represent the velocity at the end of the interval t, and let $o'r$, or the equal and parallel line or_1, represent the velocity at the beginning of the interval t. Since the velocity is continually changing in direction, a force f must continually act, and since the direction of this force is always at right angles to the velocity (see preceding paragraph), it must act continually toward the center. Its value is then

Uniform circular motion.

$$f = ma = m\left(\frac{r_1r_2}{t}\right)_{t \doteq 0}. \qquad (112)$$

But since S represents the constant speed and t the element of time considered, it is evident that

$$o'o = St.$$

But $$\frac{o'o}{R} = \theta, \quad \text{and} \quad \frac{r_1r_2}{S} = \theta.$$

Hence $$r_1r_2 = \theta\, S = \frac{S^2t}{R}.$$

Substitution of this value in (112) gives

$$f = \frac{mS^2}{R}. \qquad (113)$$

But if ω represent the *angular* speed, then $\omega = \frac{S}{R}$.

Hence
$$f = \frac{mS^2}{R} = m\omega^2 R, \tag{114}$$
an equation which asserts that the central force which must be applied to keep a body in a circular orbit is directly proportional both to the second power of the angular velocity, and to the first power of the radius of the orbit.

Experiment

Object. To verify the law of centripetal force.

Method. The masses m_1 which slip along the rod *ab* (Fig. 62) are attached by cords which pass over pulleys in the case *p* to the sliding collar *c*. The central force which is necessary to keep the weights moving in a circle is represented by the tension in the cords. For a certain critical value of the speed this tension is equal to the weight of the collar *c*. In order, therefore, to verify the law stated in (114), it is only necessary to measure the radius R, the masses m_1 and *c*, and to observe the speed required to lift *c*. All forces must of course be expressed in dynes, all masses in grams, all linear distances in centimeters, all angular distances in radians.

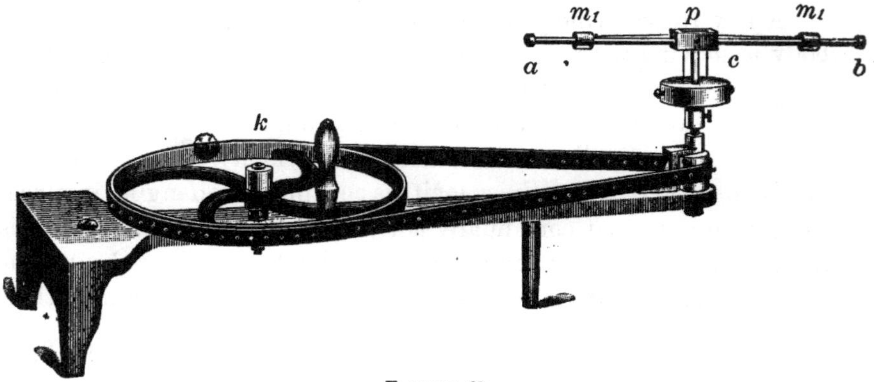

FIGURE 62

Directions. First count the number of revolutions of the axle to one revolution of the wheel *k*. Then, from measurements upon m_1, *c*, and *R*, calculate what number of turns *N* of the wheel per second is necessary just to lift *c*.

To obtain an experimental value of the same quantity, let one experimenter maintain a constant rotation of *k* at such speed that the collar *c* is either held balanced between the upper and lower stops, or else continually oscillates back and forth between them (the stops are so arranged that *c* is free to move through, only about 1 mm.). Then, as soon as this constant condition is attained, let a second experimenter take with a stop-watch the time of fifty revolutions, repeating several times in order to test the accuracy of the observations. Then change both *m* and *R* and repeat. Compare in each case the observed and calculated values of the speed.

Record

1st value of m_1 = ——		R = ——	c = ——	∴ N calc. = ——
N obs. 1st trial = ——	2d = ——	3d = ——	mean = ——	% error = ——
2d value of m_1 = ——		R = ——	c = ——	∴ N calc. = ——
N obs. 1st trial = ——	2d = ——	3d = ——	mean = ——	% error = ——
3d value of m_1 = ——		R = ——	c = ——	∴ N calc. = ——
N obs. 1st trial = ——	2d = ——	3d = ——	mean = ——	% error = ——

Problems

1. Taking the radius of the earth as 6370 kilometers, find how many dynes of force are required to hold a gram of mass upon the surface (1) at the equator; (2) in latitude 45°. Hence, find what would be the values of *g* at the equator and in latitude 45° if the earth did not rotate. Also find how many times the velocity of rotation would need to be increased in order that bodies at the equator might have no weight.

2. The radius of the moon's orbit is approximately 60 times the radius of the earth. Calculate the force in dynes which must act upon each gram of the moon's mass in order to hold it in its orbit, the period of the moon's rotation being 27 days 8 hours. Compare this result with the force of the earth's attraction upon a gram of mass at the distance of the moon as computed from the law of gravitation. It was precisely this computation which led Newton to assert the law of gravitation.

3. A skater describes a circle of 10 meters radius with a speed of 5 m. per second. What must be his angle of inclination to the vertical in order that he may be in equilibrium?

MOLECULAR PHYSICS AND HEAT

XIV

BOYLE'S LAW

Theory

The elastic properties of gases were very early made the subject of observation and speculation, but the first results of experiments made for the purpose of discovering the exact relation which exists between the pressure exerted by a confined gas and the volume which it occupies, were published by the English physicist Boyle in 1661 in a work entitled "Defence of the Doctrine of the Spring and Weight of Air." These experiments brought to light the law which has since been called Boyle's Law; sometimes also called Mariotte's Law. This law asserts that so long as the temperature remains constant, the pressure which a gas exerts upon the walls of the containing vessel is directly proportional to its density or inversely proportional to the volume which it occupies; symbolically,

Discovery of Boyle's Law.

$$\frac{P_1}{P_2} = \frac{d_1}{d_2} = \frac{V_2}{V_1}; \tag{115}$$

or

$$P_1 V_1 = P_2 V_2 = P_3 V_3 = \text{etc.} = \text{constant}. \tag{116}$$

The French physicist Mariotte independently discovered the same law fifteen years later.

Before the discovery of Boyle's Law, in fact before the beginning of the Christian era, two theories had been advanced to account for the elastic properties of air. The first was the repulsion theory, according to which the pressure exerted by confined air was attributed to repellent forces existing between the molecules which were assumed to be at rest. This theory was held by prominent scientists even as late as the middle of the nineteenth century. In order to reconcile the theory with Boyle's Law, it is necessary to assume that the molecules repel each other with forces which are inversely proportional to the distances between them. The theory has now been altogether abandoned; first, because such a law of molecular force is wholly at variance with all modern views as to the nature of molecular force; second, because it necessitates the conclusion that the pressure which a gas exerts is a function not of its density and temperature alone, but also of the shape and size of the containing vessel, a conclusion which is directly contradicted by experiment; third, because the fact that a gas does not experience a rise in temperature when it expands into a vacuum, proves that no repulsion exists between its molecules.

The repulsion theory of gases.

According to the kinetic theory, the pressure which a gas exerts against the walls of a containing vessel is due to the bombardment of the walls by rapidly moving molecules which at ordinary pressures are so far apart that they exert no forces whatever upon one another, and which occupy so little space themselves that the total number of impacts per second against the walls is simply the product of the number of molecules present and the number of impacts which one single molecule would make if it were alone in the vessel and were moving with the mean velocity of all the molecules. Although a crude form of this theory is as old as Greek philosophy, it can not be said to have taken definite shape before about 1738, when it was advanced by Daniel Bernouilli. It did not gain general acceptance until the middle of the nineteenth century, when the labors of Joule in England and of Clausius in Germany won for it well-nigh universal support.

The kinetic theory of gases.

While the repulsion theory was unable to account for Boyle's Law without the aid of a highly improbable assumption, the kinetic theory furnishes an immediate explanation of this law. For, manifestly, if gas pressure is due to impacts alone, its value at any instant must be the product of the mean force of each impact and the number of impacts taking place at that instant upon a square centimeter of surface. For a given gas, the first factor would depend simply upon the mean velocity of the molecules. For a constant mean velocity, the second factor would be proportional to the number of molecules present in a cubic centimeter; i.e., to the density. If, then, the constancy of the mean molecular velocity be taken as the condition of constant temperature, it follows at once from the kinetic theory, that the pressure should be directly proportional to the density. This is Boyle's Law.

Up to 1848 Boyle's Law was supposed to hold rigorously for the so-called permanent gases. In this year, however, the French physicist Regnault performed very careful experiments which showed that for air at ten atmospheres pressure, the product $P V$ differs by about one-fourth of one per cent from its value at one atmosphere. At higher pressures, the departure is more marked, amounting at 600 atmospheres to more than 25 per cent. These departures are evidence for, rather than against, the correctness of the kinetic theory; for when the molecules are crowded so close together that the space which they themselves occupy is no longer negligible in comparison with the total volume of the vessel, then the kinetic theory would require that the pressure increase *more* rapidly than the density; i.e., that the product $P V$ increase. This is what actually occurs in the case of all gases when the pressures are very high (100 atmospheres or more). For moderate pressures (1 to 50 atmospheres), the departures are in the opposite direction in the case of all gases excepting hydrogen (and probably also helium); i.e., the pressure increases *less* rapidly than the density, or, in other words, $P V$ decreases. This is due to the fact that the attractive forces between the molecules are not wholly negligible in comparison with the forces of impact.

Departures from Boyle's Law.

Experiment

Object. To verify Boyle's Law for ordinary pressures.

The body of dry air which is to be experimented upon is contained in the upper part of the graduated tube *a* (see Fig. 63), which has a diameter of about 1 cm. The lower end of this tube is beneath the mercury in the cistern *AB*. The air-jacket which surrounds *a* serves to maintain a constant temperature throughout the experiment. If it is desired to take

Method.

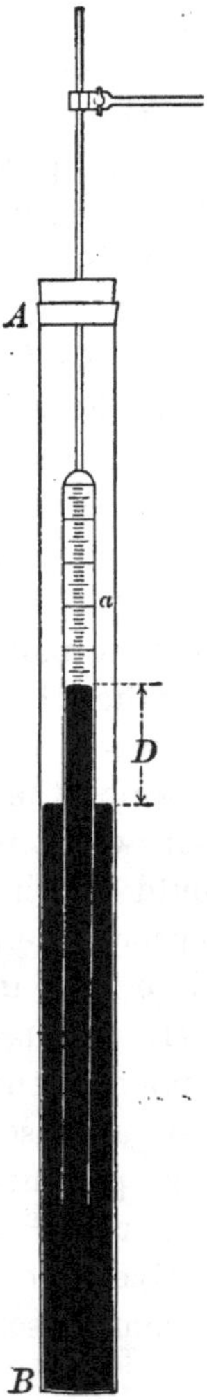

FIGURE 63

still further precautions, the upper part of AB may be filled with water, although this is generally found to be unnecessary. The tube a is graduated in cc., so that the volume V may be obtained directly by reading the scale upon a. The determination of the pressure P, which corresponds to any particular volume V, requires the observation, first, of the barometer height H, and second, of the difference in level D between the mercury in AB and in a. This observation is made by means of the cathetometer (see below). It is evident, therefore, that if the temperature of the mercury in the barometer is the same as that of the mercury in AB, then P, expressed in centimeters of mercury, is equal to $(H-D)$. If the two temperatures are different, H must be corrected by multiplying the observed height by the ratio of the density of mercury at the temperature of the barometer and its density at the temperature of AB. This correction is wholly negligible in this experiment unless the difference in the two temperatures amounts to more than 5°C.

DIRECTIONS.—To determine the barometric height H, proceed as follows: By means of the thumbscrew s [see Fig. 64, (1)], raise or lower the level of the mercury in the cistern E of the barometer until the ivory point n just touches the mercury surface. This setting can be made with great accuracy by observing when the image of the point which is seen in the mercury, just appears to come into contact with the point itself. This point n is the zero of the

Reading of the barometer.

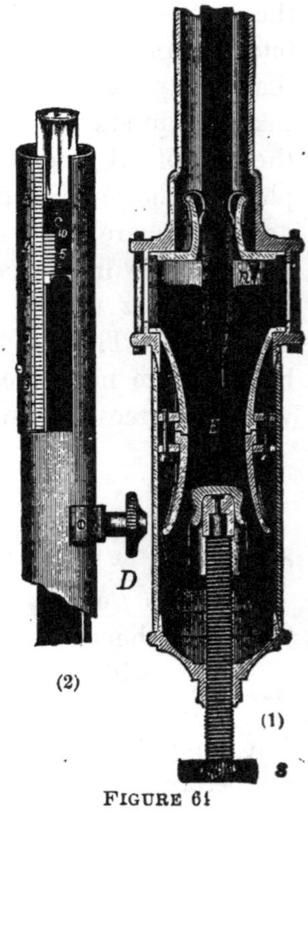

FIGURE 64

scale which is attached to the upper portion of the metal case surrounding the barometer tube. By turning the milled head D, move the vernier c with which the scale is provided until its lower end is clearly above the convex surface of the mercury. Then carefully lower it until it appears to be just in contact with the highest point of this convex surface. During this operation, keep the eye in such a position that the back lower edge of the vernier tube seems to coincide with the front lower edge. To test the setting, move the head up and down, and see to it that the white background behind the barometer never becomes visible above the top point of the meniscus. Read the scale and vernier (see Appendix for theory of vernier). The capillary correction for the barometer is to be obtained from the Appendix table, which is headed "Capillary Depression of Mercury."

Adjust the cathetometer (see Fig. 65) in three steps, as follows:

Adjustment of the cathetometer.

(1) *To make the column vertical*, loosen the set screw s_2 and turn the column until the telescope is at *right angles* to the line connecting two of the feet; e.g., A and B. Then bring the bubble to the middle by means of the leveling screw in the third foot C. Next rotate the column through 180° about its own axis, and if, after rotation, the bubble is displaced from the middle, correct half of the angular error by means of the leveling screw in the foot C and the other half by means of the screw s which inclines the telescope. If the bubble is against one end of the level-tube, it is impossible to know when just half of the angular correction has been made. Hence it is necessary first to estimate roughly these half-corrections, then to rotate again through 180°, and

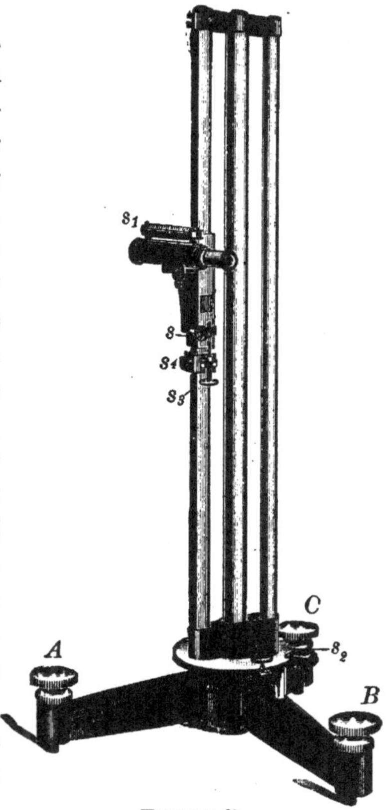

FIGURE 65

again to correct, and thus to proceed until the bubble remains, upon reversal, somewhere near the middle of the tube. The half-corrections can then be read off accurately upon the scale on the level-tube. When this adjustment has been made to such a degree of accuracy that a reversal displaces the bubble through perhaps one or two divisions, turn the column this time through 90°, i.e., until the level is *parallel* to the line connecting *A* and *B*, and bring the bubble back to the middle by turning the leveling screws *A* and *B* equal amounts in opposite directions. The column should now be approximately vertical. In order to make it accurately vertical, the whole operation must be repeated from the beginning with more care. After the completion of the second leveling, rotation of the column into any position whatever should not cause a displacement of the bubble of more than half of a division.

(2) *To make the line of sight coincident with the axis of the telescope*, first focus the eye-piece carefully upon the cross-hairs by slipping the former forward or back in the draw-tube; then, by means of the rack and pinion with which the draw-tube is provided, focus the telescope sharply upon the scale on tube *a* (see Fig. 63), which should be set at a distance of about a meter from the cathetometer. If moving the eye slightly from side to side causes the cross-hairs to appear to move at all with reference to the scale, repeat both of these focusings until this parallax effect is wholly removed. Now turn the telescope in its socket until one of the cross-hairs is parallel to the scale divisions, and by means of the screw s_3, which moves the whole telescope up or down, set this cross-hair upon some chosen division of the scale; then rotate the telescope in its socket, i.e., about its own axis, through 180°. If this operation changes at all the reading of the cross-hair upon the scale, correct half of the error by means of s_3 and the other half by means of the small screw which is found at one side of the eye-piece and which moves the cross-hair across the field of view. Repeat this adjustment until rotation of the telescope through 180° produces no change in the reading. The line of sight then coincides with the axis of the telescope.

(3) *To make the axis of the telescope horizontal*, proceed in either one of the following ways [method (b) is generally to be recommended]: (a) Bring the bubble to the middle by means of

screw s, then lift the level carefully from the telescope-tube, turn it end for end and replace. If this operation displaces the bubble, correct half of the error by means of s, the other half by means of the screw s_1, which adjusts the position of the level-tube in its case. The telescope-tube, and hence also the line of sight, will be horizontal when no displacement of the bubble is produced by a reversal of the level. (b) Set the cross-hairs of the telescope upon some point on a scale about a meter distant. Then take the *telescope* out of its socket, turn it end for end and replace. Next rotate the vertical column through 180° and look again at the chosen point. If the cross-hairs are no longer upon it, correct half the displacement by means of the screw s which inclines the telescope, and half by means of the screw s_3 which raises it vertically. The telescope-tube will be horizontal when, after reversal and rotation, the same point of the scale comes under the cross-hair.

Reading the mercury levels.

When the cathetometer is in complete adjustment, carefully loosen the set screw s_4, slide the telescope up or down the column until the cross-hair is near the top of the meniscus of the mercury in the cistern AB (see Fig. 63), clamp s_4 and make the final setting upon the meniscus by means of the fine adjustment screw s_3. Then take the reading of the vernier upon the scale of the cathetometer column. Next raise the telescope, set the cross-hair upon the top of the mercury meniscus in the tube a, and take a second reading. The difference between the two readings gives the distance D. Take a number of observations of this height in order to see to what degree of accuracy it is obtainable. At the same time read through the telescope the volume V upon the scale on tube a. This reading should not be taken either at the top or at the bottom of the meniscus, but at such an intermediate point as would correspond to the same volume if the meniscus were flat. This point can be obtained only by careful estimate. Starting with a pressure which is but a trifle less than one atmosphere, vary the volume by five about equal steps until it is as large as can be conveniently obtained with the apparatus, and take the five corresponding pressure readings. If the barometric height varies during the experiment, take this fact into account.

Record

V	H	Read'gs 1st trial			Read'gs 2d trial			Mean D	$\therefore P$	$\therefore PV$	% dif. from mean
		in AB	in a	$\therefore D$	in AB	in a	$\therefore D$				

Problems

1. Make estimates of the probable observational errors in both P and V above, and thence deduce the maximum permissible error in PV. Compare with last column of the record.

2. A level is in adjustment when the line joining the points upon which it rests is parallel to the tangent drawn to the highest point of the level-tube, i.e., when the line cd (see Fig. 66) is parallel to the line ab. Show why in adjusting a level which is out of adjustment and in leveling the table upon which it rests (see Fig.) it is necessary after reversal to correct half at s and half at s'. Hence justify throughout the methods used in adjusting the cathetometer.

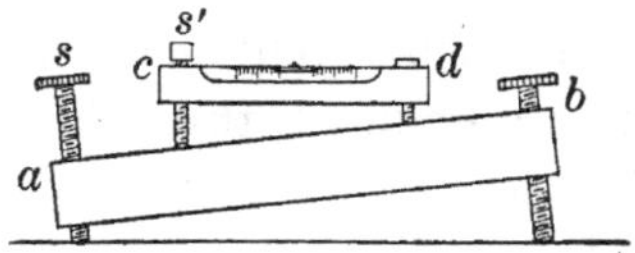

FIGURE 66

3. A confined body of air V is placed under a pressure of P mm. of mercury (see Fig. 67. $P = H + ab$). By means of the three-way stopcock s, the connection between the two arms (see Fig.) is then shut off and a volume v of mercury drawn from the right arm. The level of the mercury in this arm then sinks to the point a'. Connection between the arms is then reëstablished by means of the cock and the left arm lowered until the level in the right arm is again at a'. The pressure upon the air in the bulb is now found to be h mm. less than at first. Find V, the

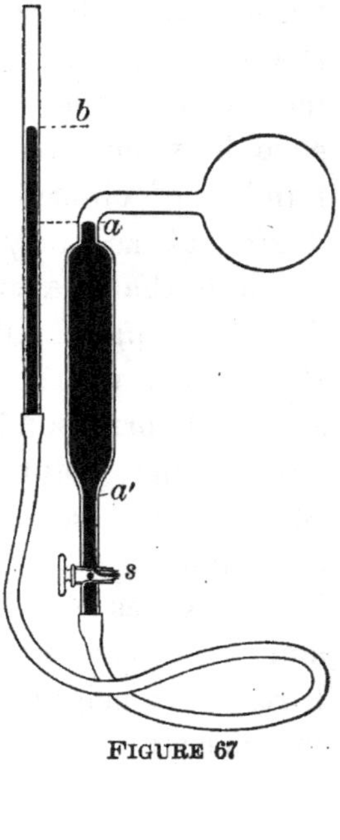

FIGURE 67

volume of the bulb down to the point a, in terms of v, P, and h.

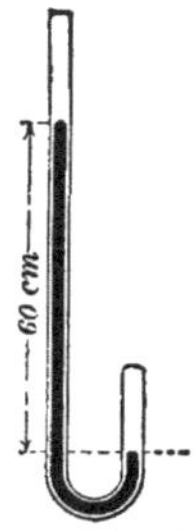

4. In Problem **3**, V was found to be 500 cc. A powder was introduced into V and the mercury in the right arm brought back to its original height a. The pressure was then found to be 900 mm. 250 cc. of mercury were drawn off at s precisely as above, and the pressure was found to fall to 500 mm. Find the volume of the powder.

FIGURE 68

5. A volume of 30 cc. of air is confined in the closed arm of a manometer (see Fig. 68). In the open arm the mercury stands 60 cm. higher than in the closed arm. What will be the difference in the levels when the volume of the air is reduced to 10 cc.? (Bar. Ht. = 76 cm.)

XV

DENSITY OF AIR

Theory

Density of air from two pressures, a weight and a volume.

The existence of Boyle's Law makes the determination of the density of air at a given temperature a very simple matter, at least in theory. For, let a globe A of known volume V, containing air under atmospheric pressure P_1, be balanced upon a beam of equal arms against some body W (see Fig. 69). Then let air be forced into the globe until the pressure within it is P_2. Suppose that the weight now needed to balance the globe is $W + w$. Since it requires the same weight to balance the glass in each case, it is evident that w is the weight of the air which has been forced into the globe in changing the pressure from P_1 to P_2. If, then, V represent the volume of the globe, and d_1 and d_2 the densities of air corresponding to the pressures P_1 and P_2 respectively, it is evident that

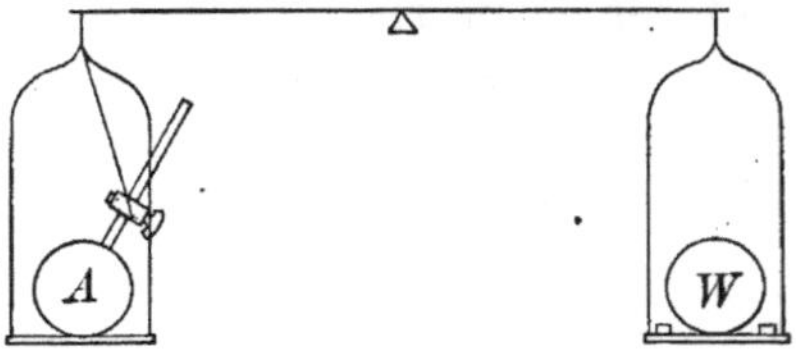

FIGURE 69

$$Vd_2 - Vd_1 = w. \tag{117}$$

But by Boyle's Law,

$$\frac{d_1}{d_2} = \frac{P_1}{P_2}. \tag{118}$$

The elimination of d_2 from (117) and (118) gives the density which is sought, viz.,

$$d_1 = \frac{wP_1}{V\,(P_2 - P_1)}. \tag{119}$$

If the counterpoise W consists of a few small weights and a closed glass globe of the same external volume as A, any errors which would arise from changes in the barometric height or the temperature during the experiment are altogether eliminated, provided only that the temperature at which the air was introduced into the bulb under pressure P_2 is the same as that which corresponds to P_1. For, with this arrangement, the buoyant effect of the air upon both sides of the balance is the same no matter how rapidly the barometric pressure or the temperature may change. This device was first used by Regnault in his classical determination of the density of air.

Experiment

Object. To find the density of dry air under existing conditions of temperature and pressure.

DIRECTIONS.—1. In order to obtain dry air for the experiment, first see to it that the stopcock c is well greased, then connect the bulb to an air pump through a calcium chloride drying-tube in the manner shown in Fig. 70. A pump capable of produc-

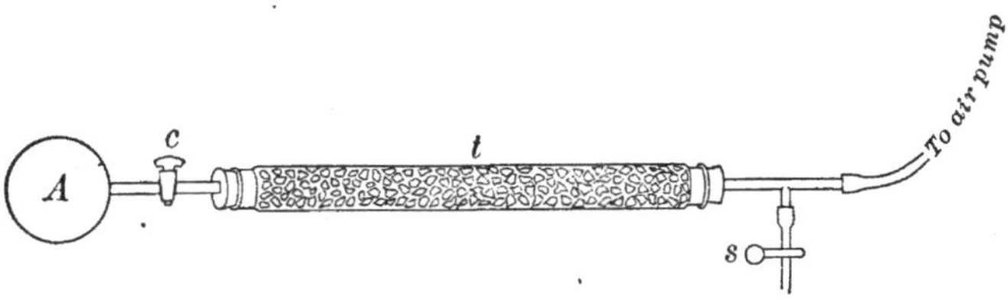

FIGURE 70

ing a high vacuum is unnecessary. The water pump of Fig. 87 may be conveniently used. First close the pinchcock s and exhaust. Then very slowly open s, and thus fill A with air which has been dried by passage through the calcium chloride tube. Repeat this operation from two to six times, according to the degree of exhaustion obtainable with the pump. Then, in order that the bulb may assume accurately the outside temperature, wait about five minutes before removing the bulb from the drying-tube or closing the cock c. Observe the temperature by means of a thermometer hung near the bulb, then carefully close cock c, remove and weigh as follows:

To dry the air in the bulb.

2. First remove all dust from the bulb and the pans of the analytical balance (see Fig. 71) by means of the camel's-hair brush which will be found in the balance case. Then very carefully suspend the bulb from one of the hooks c and place upon

the other pan the counterpoise and a weight of a few grams from the box of weights, taking pains to touch these weights only with the pincers, never with the fingers. Next release the pans by pushing in and fastening the button *e* which controls the pan-arrest; then turn very slowly the milled head *a* and thus lower the beam-arrest just enough to see whether the pointer begins to move to left or right; i.e., whether the chosen weight is too heavy or too light. This done, raise the beam-arrest immediately, but so slowly as not to endanger the knife-edges by the slightest jar; replace the chosen weight by the one next heavier or next lighter, and try it in the same way. Take the utmost care *never to place a weight on the pans or to take a weight off except when the beam is arrested.* Proceeding thus, make a systematic trial of the gram weights until you know between what two consecutive numbers of grams the condition of balance must lie; then try in the same way the milligram weights in order of magnitude until a weight is found such that when the beam-arrest is completely lowered the pointer oscillates *near the middle* of the scale over from 3 to 6 divisions. A larger swing than this indicates insufficient care in lowering the arrest. The rider *r* may be used in place of the small milligram weights, if desired, but no attempt should ever be made to add fractional portions of a milligram by means of the rider. Before taking the resting point, raise again the beam-arrest, taking pains to avoid a jar by raising at a time when the pointer is at the middle of its swing, close the face of the balance case so as to shut out all air currents, and stop all swinging of the pans by alternately pressing and releasing the button *e* which controls the pan-arrest. Then carefully lower first the pan-arrest, then the beam-arrest, and take the resting point R_1. This is to be done by averaging the mean of three successive turning points of the pointer on one side with the mean of the two intervening turning points on the other

To make the first weighing.

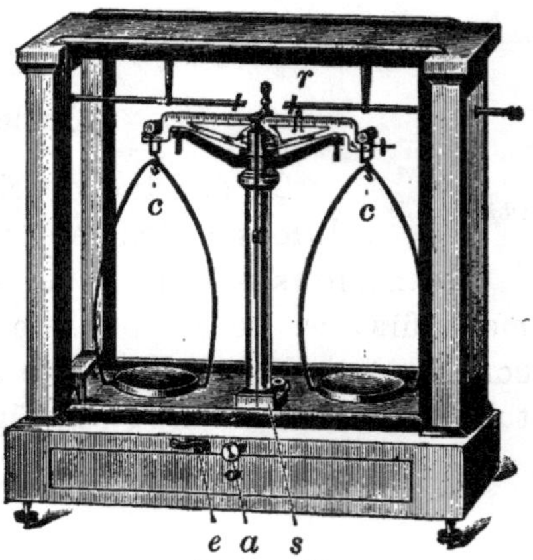

FIGURE 71

side. This use of an odd instead of an even number of turning points eliminates completely the effect of damping. The following example taken in connection with Fig. 72, which represents the scale *s* of Fig. 71, will make clear the method of procedure:

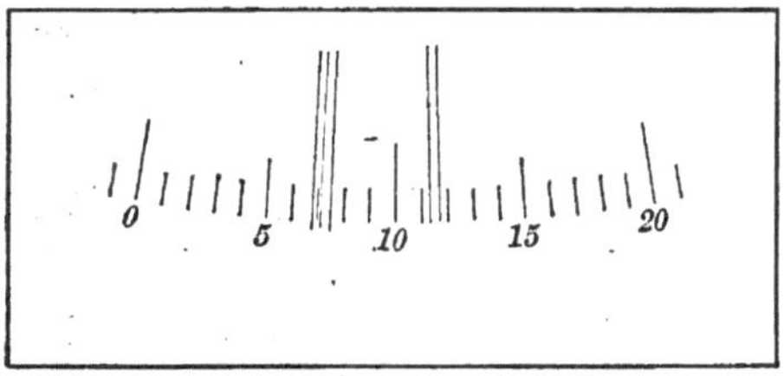

FIGURE 72

	TURNING POINTS	
	Left	Right
	6.8	
	7.1	11.2
	7.4	11.7
Means	7.10	11.45
R_1 =	9.28	

Having determined the resting point R_1, slowly raise the beam-arrest when the pointer is in the middle of a swing, then from the absences in the box of weights, count the weights which have been used and check by counting again as the weights are replaced. Call the sum of these weights W_1. Finally, close the balance-case, see that every weight is in its proper compartment in the box of weights, replace the pincers in the box, and place the latter in the drawer of the balance-case.

To fill the bulb with air under pressure P_2.

3. Fill the bulb with air under pressure P_2 as follows: Attach it securely to the pressure apparatus shown in Fig. 73. This consists of a bicycle pump *p* attached to an air-tight jar *J*, which is furnished with a mercury pressure-gauge *g* and a calcium chloride drying-tube *t*. Open communication between the jar and the bulb through the drying-tube, and produce a difference of level of 50 or 60 cm. in the manometer arms. After waiting about five minutes for the compressed air to regain the tempera-

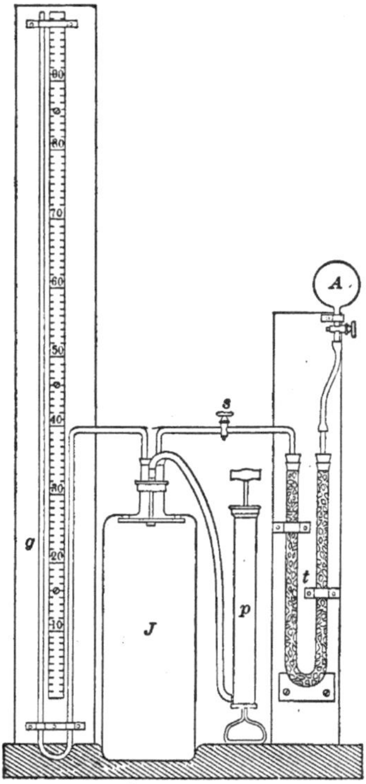

FIGURE 73

ture of the room, read simultaneously the two arms of the manometer, and at the same instant close the stopcock c. Take the temperature by means of a thermometer hung near the bulb. If this temperature differs by more than half a degree from that previously taken, the first weighing must be discarded and another taken under the same conditions of temperature as those which here exist. This need not be done until after the completion of **4**.

To make the second weighing.

4. To make the final weighing, place the bulb upon the *same scale-pan* as before and, proceeding exactly as in **2**, balance it by means of such weights W_2 that the pointer again oscillates near the middle of the scale. Then take the new resting point R_2 in precisely the manner described above. If R_2 coincided exactly with R_1, then evidently $W_2 - W_1$ would be the weight w which is sought. But, in general, R_2 will not coincide with R_1, and it is a very slovenly proceeding to attempt to make it do so, as is often done, by repeatedly shifting the position of the rider. The most rapid and the only correct method of making an accurate weighing is to determine the *sensitiveness*,* i.e., *the number of scale divisions which the pointer is shifted by the addition of one milligram*, and then to calculate by interpolation the exact correction which must be applied to W_2 in order to bring R_2 precisely into coincidence with R_1; this is done as follows: Immediately after finding R_2, add to the *lighter* side a small weight, say 2 mg. (1 mg. if the balance is very sensitive; this may be done by means of the rider if desired), and take the corresponding resting point R_3. This procedure simply determines the value in milligrams of the scale divisions. Thus, if $R_2 = 10.63$, and if, upon the addition of 2 mg., $R_3 = 7.01$, then the sensitiveness is $\frac{10.63 - 7.01}{2} = 1.81$. From the two resting points R_1 and R_2, and the sensitiveness S, it is a very easy matter

*Since, on account of the bending of the balance arms, the sensitiveness varies with the load, theory requires either that it be determined at each weighing, or else that a table showing the variation of the sensitiveness with the load be made out once for all and kept for use with the balance. Since, however, with good balances, it generally requires a very considerable change in load to produce an appreciable change in the sensitiveness, it is usually unnecessary to make more than one careful determination of the sensitiveness so long as the loads involved are of about the same magnitude.

to calculate the weight which would need to be added to or subtracted from W_2 in order that the pointer might be brought exactly to the original resting point R_1. Thus, in this case, the number of milligrams which would be required to move the pointer from R_2 [= 10.63] back to R_1 [= 9.28] is $\frac{10.63 - 9.28}{1.81} = .74$. This number of milligrams must be added to or subtracted from W_2, according as the point 10.63 is farther from or nearer to the bulb than the point 9.28. Let W_3 represent the corrected value of W_2. Then $W_3 - W_1$ is the w of equations (117) and (119).

To find the volume V.

5. To find the volume of the bulb, fill it either with water or with mercury, and weigh upon the trip-scales. Take the temperature of the liquid used and obtain the density from a table (see Appendix). From the weight and density of the liquid calculate V. If the liquid used be mercury, the filling can most easily be done by means of a funnel made by drawing down one end of a piece of glass tubing so as to form a capillary tube long enough to reach into the interior of the bulb. When the bulb is full of mercury, it must of course be handled with extreme care, the stopcock being always left open so as to prevent breakage by expansion. Filling with water may be done in the same way if sufficient pressure be applied to force the water through the capillary tube. It may also be done by alternately heating and cooling the air in the bulb, the neck being kept under water during the cooling. If water be used, the bulb will not be again fit for use until it has been thoroughly dried by repeatedly exhausting after the temperature has been raised above 100° C. by carefully heating with a rapidly moving Bunsen flame.

The barometer corrections.

The pressures which occur in the numerator and denominator of (119), viz. P_1 and $P_2 - P_1$, must of course be expressed in the same units. If they are expressed in centimeters of mercury, the two columns of mercury which they represent must have the same temperature.* It is evident, then, that since $P_2 - P_1$ represents the difference in the readings of the manometer arms at the temperature of the room, the barometric reading P_1 should also correspond to the temperature of the

* A difference in temperature not exceeding 5° C. leads to a wholly inappreciable error. It may therefore be overlooked (see table of mercury densities in Appendix).

room; i.e., for use in (119) the observed barometer height needs correction only for capillarity. But in the table of "Densities of Dry Air" given in the Appendix, the pressures all represent heights of mercury columns at 0° C. Hence, in order to compare your value of d_1 with Regnault's value, which is represented by this table, the observed barometric height must be reduced to 0° C. This reduction is made by multiplying the observed height by the ratio of the densities of mercury at the room temperature and at the zero temperature. To save labor, this correction is worked out for all ordinary temperatures in the table in the Appendix entitled "Reduction of the Barometer Height to 0° C." (This table also contains a slight correction for the expansion of the brass barometer scale.)

Record

Bar. Ht. obs'd = —— corrected for capil'ty= —— Reduced to 0°= —— cm.

Temperature of air corresponding to P_1 = —— to P_2 = ——

First resting point $[R_1]$ = —— First weight $[W_1]$ = —— gm.

Second " " $[R_2]$ = —— Second " $[W_2]$ = —— gm.

Third " " $[R_3]$ = —— Sensitiveness $[S]$ = ——

∴ correction to be applied to W_2 = —— mg. ∴ W_3 = —— gm.

∴ weight of air w introduced into bulb $(W_3 - W_1)$ = —— gm.

Gauge reading, long arm = —— short arm = —— ∴ $P_2 - P_1$ = —— cm.

Wt. of bulb + Hg —— ∴ Wt. of Hg —— Tem. —— ∴ Den. of Hg ——

∴ V= —— ∴ d_1 = —— Regnault's value = —— % error = ——

Problems

1. A glass tube open at one end is 60 cm. long. The inside is covered with a soluble pigment. After a sea sounding, in which the tube was lowered vertically, open end down, the pigment was found to be dissolved to within 5 cm. of the top. The density of sea water being 1.03, find the depth of the sea.

2. A tube 100 cm. long is half filled with mercury. It is then inverted in a cistern of mercury. How high does the mercury stand in the tube above the cistern? (Bar. Ht. 76 = cm.)

3. A barometer tube was imperfectly filled. When the space above the mercury contained 10 cc., the barometer indicated 70 cm. pressure. When the space above was reduced to 5 cc. by pushing the barometer down into the cistern, it indicated 69.5 cm. What was the true barometric height?

4. Show why the true zero of a vibration which is gradually dying down because of damping, is not obtained by taking the mean of two successive turning points, one on the right and one on the left.

5. Show why the true zero is obtained by averaging the mean of two successive turning points on one side with the intervening turning point on the other.

XVI

THE MEASUREMENT OF TEMPERATURE

Theory

It is a fact of common observation that as a body grows hot it increases in volume. Quantitative measurement shows, however, that as different bodies pass through the same change of temperature, e. g., from the freezing to the boiling point of water, their expansions per cubic centimeter are widely different. In 1787, however, a Frenchman by the name of Charles announced that all gases show the same expansions as they pass between two fixed temperatures. This result was confirmed by Gay-Lussac some twenty years later, in the first series of experiments which were sufficiently exact to thoroughly justify the conclusion. The law is now most generally known by the name of Gay-Lussac, though it is also called the law of Charles. This law, like that of Boyle, has been found by later experiment to be only approximately correct. For the permanent gases, however, the departures are only slight, as will be seen from the table on p. 126. Since the different gases obey Boyle's Law with different degrees of exactness, these departures from the law of Gay-Lussac were to have been expected. In fact, the kinetic theory requires the result, established by experiment, that the gases which show the largest departures from Boyle's Law, show also the largest deviations from the mean value of the coefficient of gaseous expansion. (See CO_2 and N_2O.)

The law of Gay-Lussac.

Direct knowledge of the temperature, i.e., of the hotness, of a body is gained only through the sense of touch. But since, as a body grows hot to the touch, it also expands, this fact of expansion has been made the quantitative measure of change in temperature, even when the change is too slight to be perceived by the touch. Thus, for example, the volume of a given weight of iron is observed first at the freezing point and afterward at the boiling point of water; the increase in volume is then divided into 100 equal parts, and 1 degree rise of

Temperature.

temperature is arbitrarily defined as any temperature change which will produce an expansion of the iron equal to one of these parts. Thermometers constructed in this way from different substances do not exactly agree with one another for temperatures other than 0° and 100°, for the reason that the expansions of the different substances are not generally the same functions of the temperature. Hence, it becomes necessary to choose arbitrarily some particular substance whose expansion shall be taken as the measure of temperature change. The gases possess peculiar advantages for this purpose, first, because all gas thermometers agree with one another (see law of Gay-Lussac) and, second, because the kinetic theory gives to a degree measured upon a gas thermometer a particular physical significance (see below). For these reasons the expansion of a gas has been chosen as the measure of temperature change, and all correct determinations of temperature are now made either by means of gas thermometers or else by means of other instruments which have been standardized by comparison with gas thermometers.

Constant-Pressure gas thermometers and the expansion coefficient of gases.

Gas thermometers take two forms: (1) the *constant-pressure* form, and (2) the *constant-volume* form. The first consists of any arrangement for measuring the expansion of a gas which is kept under a constant pressure. For example, suppose the gas to be confined within the bulb B and a part of the stem cd (see Fig. 74) by means of a small mercury index i which moves without friction forward or back as the temperature of the bulb rises or falls. Let the stem be open at d, so that the confined gas is always under the condition of pressure which exists outside. *One degree of change in temperature is then defined, in the centigrade system, as any temperature change which, starting from any temperature whatever, will produce in the confined gas an increase of volume amounting to* $\frac{1}{100}$ *of the increase which takes place when the temperature of the gas passes between the freezing and the boiling points of water.* If the barometric pressure never varied, the stem of such a thermometer might easily be graduated so that the instrument would read tem-

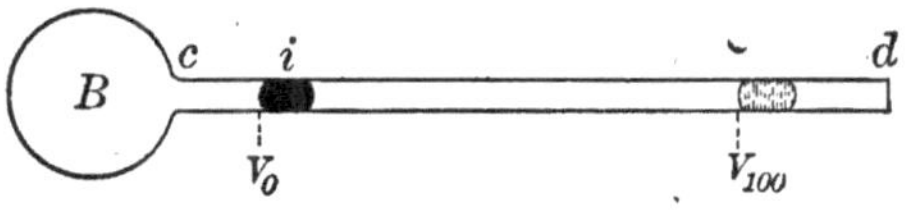

FIGURE 74

perature directly. It would only be necessary to place it horizontally, first in melting ice, then in the steam rising from boiling water; to mark the positions of the index at the two temperatures; and then to divide the increase into 100 equal parts. In practice a gas thermometer is never treated in this way, for the reason that such a graduation would give correct temperature readings only for the particular pressure P (e.g., 76 cm.) at which the graduation was made. In general, then, in order to make a correct determination of any unknown temperature with such a thermometer, it is necessary to know, not only the index reading when the bulb is at the unknown temperature t, but also the index readings which correspond, *at the existing pressure*, to the freezing and boiling temperatures. From these three readings the temperature t is obtained without any actual determination of any one of the three volumes, i.e., the volume at 0° [$=V_0$], the volume at 100° [$=V_{100}$], or the volume at t° [$=V_t$]; for it is only the volume differences ($V_t - V_0$) and ($V_{100} - V_0$) which need be known. Thus it is at once evident from the definition of temperature given above, that from such observations the temperature on the centigrade scale is given by

$$t = \frac{V_t - V_0}{\frac{V_{100} - V_0}{100}}. \tag{120}$$

The observation at 100° can be dispensed with if the actual volume V_0 be determined, and if the coefficient of expansion of the gas between 0° and 100° be known. For, the coefficient of expansion α between any two temperatures t_1 and t_2 is defined as the increase in volume corresponding to a rise in temperature from t_1 to t_2, expressed in terms of the volume at 0°. Thus the following equation is arbitrarily taken as the definition of α between t_1 and t_2:

$$\alpha = \frac{V_{t_2} - V_{t_1}}{(t_2 - t_1)\,V_0}. \tag{121}$$

The coefficient of expansion between 0° and 100° is then evidently given by

$$\alpha = \frac{V_{100} - V_0}{100\,V_0}. \tag{122}$$

Now, if this equation be combined with (120), there results

$$t = \frac{V_t - V_0}{a V_0}. \qquad (123)$$

If, then, a be known, it is evident from (123) that a determination of the temperature t may be made by means of the constant-pressure gas thermometer, by measuring the volume of the gas at 0°, and the increase in volume between 0° and t°. Either of the characteristic equations of a constant-pressure gas thermometer, (120) or (123), may be taken as the *definition* of temperature on the centigrade scale.

Constant-volume air thermometers and the pressure coefficient of gases.

Now suppose that the confined air in the bulb B, after being raised at constant pressure P_0 from the freezing to the boiling point of water, i.e., from V_0 to V_{100}, had been brought back again, while at the boiling temperature, from V_{100} to V_0 (see Fig. 74) by the application of additional outside pressure, the final value of which is represented by P_1. Then Boyle's Law gives

$$\frac{V_{100}}{V_0} = \frac{P_1}{P_0}.$$

If this equation be compared with (122), which defines the coefficient of expansion a, it is seen that

$$a = \frac{V_{100} - V_0}{100 V_0} = \frac{P_1 - P_0}{100 P_0}. \qquad (124)$$

Now, since, in the whole operation which has been considered, the gas has been brought back to its original volume, and the only *resultant* change which has taken place is the change in temperature from 0° to 100°, it is clear that P_1 is the final pressure which the gas would have exerted if its volume had always remained constant at V_0, while its temperature was raised from the freezing to the boiling point of water, i.e., P_1 is the pressure which the gas in a *constant-volume thermometer* (see Fig. 75) would have assumed at 100°, if its pressure at 0° was P_0. Call it henceforth P_{100} and write (124),

$$a = \frac{P_{100} - P_0}{100 P_0}. \qquad (125)$$

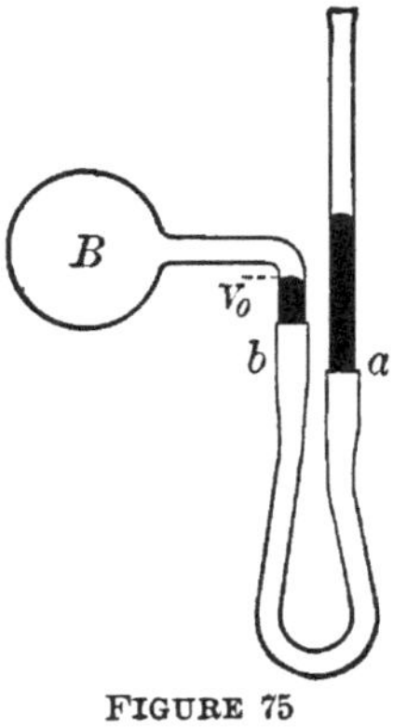

FIGURE 75

Also conceive the above operation to be replaced by that of raising the arm a (Fig. 75) and holding the mercury in b always at the same point V_0 while the temperature of B is raised from 0° to 100°, an operation which corresponds to that usually performed in the use of a constant-volume thermometer. Now the expres-

, applied to a constant-volume thermometer, is

usually called the *pressure coefficient* of the gas, and is denoted by the letter β. It thus appears that *for a gas which follows Boyle's Law the expansion coefficient* a *and the pressure coefficient* β *are identical quantities, and hence that the equation*

$$\frac{P_t - P_0}{\beta P_0}, \qquad (126)$$

applied to a constant-volume gas thermometer, gives precisely the same definition of temperature as does (123) *applied to a constant-pressure gas thermometer.*

In practice, the constant-volume thermometer is the more convenient, as well as the more accurate instrument. It is therefore almost exclusively used. But there are other reasons aside from those of convenience and accuracy, which make a choice necessary. As has just been proved, the pressure and expansion coefficients *must* be identical for gases which rigorously follow Boyle's Law. But, since such gases do not exist, it was to have been expected that refined measurements would show slight differences between these coefficients. The following table gives the results of some of the best determinations, those of Chappuis at the International Bureau of Weights and Measures being probably the most accurate yet made

Gas	Expansion Coefficient a	Pressure Coefficient β	Observer	Date
Hydrogen............	.0036600 *	.0036625	P. Chappuis	1887
Hydrogen............	.003661	003667	Regnault	1840
Air..................	.003670	.003665	Regnault	1840
Air..................		003663	Kuenen and Randall	1895
Carbon dioxide (CO_2).	.003710	.003688	Regnault	1840
Nitrous oxide (N_2O)..	.003719	003676	Regnault	1840
Oxygen		003674	von Jolly	1874
Nitrogen .		003667	von Jolly	1874
Argon		003668	Kuenen and Randall	1895
Helium...		003665	Kuenen and Randall	1895

* See Travers' "Experimental Study of Gases," p, 149. Macmillan & Co 1901.

Because, then, of the slight difference in the two coefficients for the same gas, and because of the slight differences existing between the different gases as regards the same coefficient, *the International Committee of Weights and Measures adopted in 1887 the constant-volume hydrogen thermometer as the standard of temperature measurement.* Hence, the above approximately correct definition of temperature must be replaced by that embodied in (126) when applied to a *hydrogen* thermometer. In other words, since $\beta = .0036625 = \frac{1}{273}$, *1° C. is by definition such a temperature change as will cause the pressure of a body of hydrogen which is kept at constant volume to change by* $\frac{1}{273}$ *of its zero value.*

The kinetic theory and temperature.

According to the kinetic theory, the pressure exerted by a gas which follows Boyle's Law is, at any instant, the product of the mean force of each impact and the number of impacts occurring at that instant upon each sq. cm. of surface (see Ex. XIV). Now, the number of impacts which a single molecule which is moving within a given enclosure will make per second upon the walls is evidently proportional to its velocity. Hence, with a fixed number of molecules present in the enclosure, the number of impacts occurring at any instant must be directly proportional to the average velocity. But the mean force of each impact is also proportional to the velocity (see second paragraph below). Hence with a given kind of molecule, i.e., a given gas, the pressure exerted against the walls is proportional to the *square* of the molecular velocity. But also, for a given kind of molecule, the kinetic energy ($\frac{1}{2}mv^2$) of each molecule is proportional to the square of the velocity. Hence the ratio of the two pressures which a constant volume of gas exerts at different temperatures is simply the ratio of the mean molecular kinetic energies. Since, then,

$$\frac{P_t}{P_0} = \frac{KE_t}{KE_0}, \tag{127}$$

it follows that the equation (126) which, when applied to a constant volume of hydrogen, defines temperature, may also be written

$$t = \frac{(KE)_t - (KE)_0}{\beta\,(KE)_0}. \tag{128}$$

Since $\beta = \frac{1}{273}$, and since, when $(KE)_t - (KE)_0 = \beta (KE)_0$, $t = 1$, this may be expressed in words thus: 1° *change in temperature means merely the addition (or subtraction) of a given amount to the mean kinetic energy of translation of the flying hydrogen molecules, and this amount is* $\frac{1}{273}$ *of the kinetic energy which these molecules possess at* 0° *C.*

The absolute zero of temperature.

It is evident, then, that if, starting from 0°C., this amount were subtracted 273 times, all kinetic energy would be removed from the hydrogen molecules; i.e., they would come completely to rest. Further, the interpretation of the law of Gay-Lussac, in the light of the kinetic theory, is that the molecules of all other gases would come to rest at the same temperature, viz., at −273°C. This point at which all molecular motion would cease is called the *absolute zero of temperature.* Of course this temperature can never be attained; but within the last twenty-five years great strides have been made toward it. Up to 1877, the lowest attained temperature was −110°C., produced by Faraday in 1845 by the rapid evaporation in vacuo of a mixture of ether and solid CO_2. The temperature of the solid CO_2 alone is −80°C., that of the mixture ordinarily about the same. In 1877 a Swiss, Pictet, and a Frenchman, Cailletet, independently liquefied oxygen, which has a boiling point of −182.5°C. But as neither of these experimenters obtained the liquid oxygen in a static condition, they could make no observation of its temperature. The lowest measured temperature was −140°C., obtained by Pictet by the rapid evaporation of liquid N_2O. Following these, the two Poles, Wroblewski and Olszewski, and the Englishman Dewar accomplished the liquefaction of air in quantity, and by its evaporation in vacuum produced and measured temperature as low as −210°C. In 1885, Professor Olszewski liquefied hydrogen and located its boiling point at −243.5°. It has since been not only liquefied in quantity, but also solidified by Dewar (1900), according to whose most recent determinations the boiling point of hydrogen, measured by means of a hydrogen thermometer, is between −252°C. and −253°C. Its melting point is between −256°C. and −257°C. The lowest temperature obtained by evaporating solid hydrogen is −260°C., only 13° above absolute zero.

Thus far it has been shown that 1° rise in temperature signifies an increase in the mean molecular energy amounting to $\frac{1}{273}$ of the zero value of this energy, but no attempt has yet been made to find the actual value of this zero energy, or to discover whether any simple relation exists between the mean kinetic energies of the molecules of different gases which have the same temperature. In 1851, the English physicist Joule worked out in the following way the precise relation which must exist between the pressure exerted by a gas and the mean kinetic energy of the individual molecules.

Pressure and molecular kinetic energy.

Let the gas be contained in the vessel mn (see Fig. 76), the lengths of whose sides are a, b, c, and let N represent the number of molecules in the vessel and v the average velocity of each molecule. The particles are, of course, moving in all possible directions, but, on the whole, the pressure must be just the same as though one-third of them were moving parallel to each of the three edges a, b, c. If a single particle were alone in the vessel and were moving back and forth parallel to one edge, e. g., to c, it would make against the upper wall $\frac{v}{2c}$ impacts per second; for v is the distance which it moves per second and $2c$ is the distance moved between successive impacts against this wall. If there are other molecules present in the vessel, the molecule considered may, of course, collide with them in its excursions to and fro, but, so long as the volume occupied by the molecules themselves is negligible in comparison with the volume of the vessel, the number of impacts against the wall ab will be unaltered by these collisions. For (see Ex. VII, Problem 1) when two equal elastic particles collide, the effect is the same as though one particle had passed through the other without influencing it.* Since, then, each particle which is moving parallel to c makes $\frac{v}{2c}$ impacts per

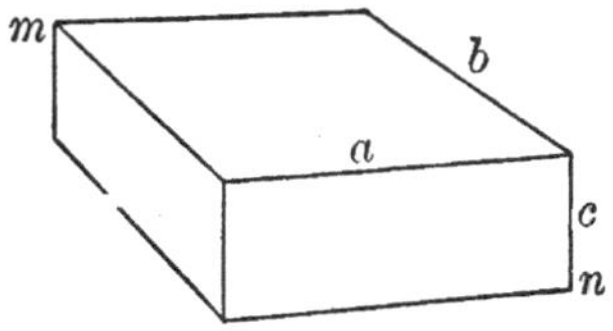

FIGURE 76

*This was proved only for the case of central impact and can not therefore be taken as a complete justification for the preceding statement. However, a more rigorous analysis than the one here given leads to precisely the same final result.

second against the side ab, and since there are $\frac{N}{3}$ particles so moving, the total number of impacts per second against ab must be $\frac{N}{3} \times \frac{v}{2c}$. Now, at each impact each molecule first loses all its velocity and then gains an equal opposite velocity. Hence the change in velocity which each molecule experiences at each impact is $2v$. The change in momentum which each particle experiences at impact is therefore $2mv$, m being the mass of each molecule. The total change in momentum experienced by the total mass which impinges per second against ab is then $2mv \times \frac{N}{3} \times \frac{v}{2c}$. But rate of change of momentum is the measure of force (see Ex. II). Hence the force f, which the wall ab experiences because of the molecular impacts, is given by

$$f = 2mv \times \frac{N}{3} \times \frac{v}{2c} = \frac{1}{3}\frac{Nmv^2}{c}. \tag{129}$$

But the pressure p is by definition the force per unit area. Hence,

$$p = \frac{1}{3}\frac{Nmv^2}{abc}. \tag{130}$$

But abc is merely the volume V of the vessel. Hence equation (130) becomes

$$pV = \frac{1}{3} Nmv^2. \tag{131}$$

Now, Boyle's Law asserts that the product of the pressure and volume of any gas is constant so long as the temperature remains constant. Hence, according to equation (131), the condition for constant temperature is a constant mean velocity of molecular motion. If n denote the number of molecules in unit volume, then $n = \frac{N}{V}$, and (131) becomes

$$p = \frac{1}{3} nmv^2. \tag{132}$$

The kinetic energy of the n molecules is $\frac{1}{2}nmv^2$. *Hence the pressure exerted by a gas is numerically equal to two-thirds of the kinetic energy of translation of the molecules in unit volume.*

Equality of kinetic energy the condition of equality of temperature. Let two gases 1 and 2 be contained in different vessels, but let them have the same temperature and exert the same pressure. According to equation (132) the pressure in vessel 1 is

$$p = \frac{1}{3} n_1 m_1 v_1^2.$$

Similarly in vessel 2,

$$p = \tfrac{1}{3} n_2 m_2 v_2^2.$$

Now, according to the law of Avogadro, the proof of which will be advanced in the next section, if the two gases exert the same pressure and are at the same temperature, they contain the same number of molecules per unit volume, i.e., in this case $n_1 = n_2$. Hence,

$$m_1 v_1^2 = m_2 v_2^2;$$

i.e., *at a given temperature the molecules of all gases have the same average kinetic energy of translation.* Thus the kinetic theory leads not only to the conclusion that 1° rise of temperature means an increase of kinetic energy amounting to $\frac{1}{273}$ of the zero energy, but also that equality of temperature in gases means equality in the average kinetic energies of the molecules.

FIGURE 77a

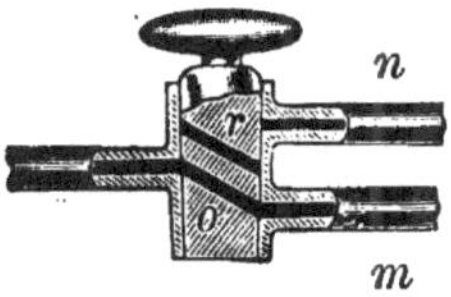

FIGURE 77b

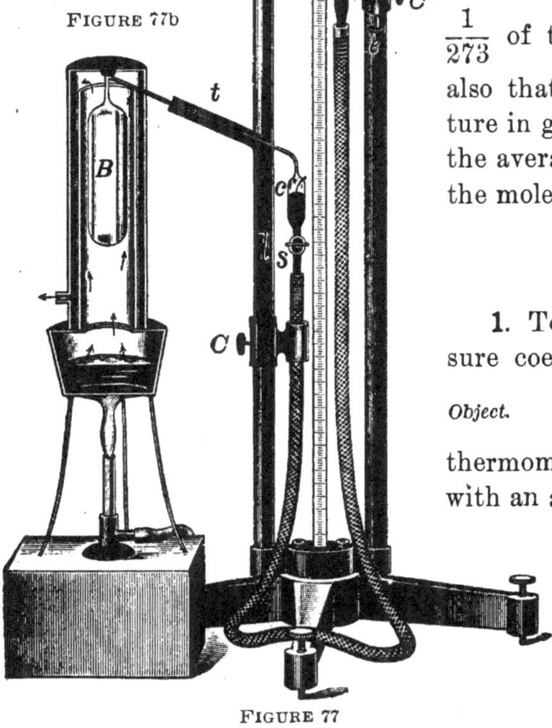

FIGURE 77

Experiment

1. To determine the pressure coefficient β of air. 2. To standardize a "mercury in glass" thermometer by comparing it with an air thermometer.

Object.

Fig. 77 shows the constant-volume air thermometer which is to be used. The volume is kept constant by so adjusting the height of the right manometer-arm that the mercury in the left arm is brought, before each reading, exactly into coincidence with the platinum point *c* (see also Fig. 77a). The fine adjustment screw *e* facilitates this setting. The difference in the mercury levels in the two arms is read off upon the central graduated scale. A mirror is attached to the back of the vernier index *i* and slides with it so that the readings can be made entirely free from the error of parallax. A three-way stopcock *s* makes it possible to put the bulb into communication either with the manometer or with the outer air. The form of this cock is shown in Fig. 77b, which represents the position when it is in communication, through the hole *o* and the tube *m*, with the manometer. Turning the cock through 180° brings the bulb into communication, through the hole *r* and the tube *n*, with the outside air. A rigid arm *l*, which is attached to the sliding collar *C*, holds the bulb in place and protects it from breakage.

Instrument.

If p_0 represent the observed pressure at 0°, H the barometric height and h_0 the difference between the readings in the two arms when the bulb is surrounded with melting ice and the mercury is brought into exact contact with *c*, then evidently

Method.

$$p_0 = H + h_0,$$

h_0 being, of course, negative if the level in the left arm is higher than that in the right. Similarly the observed pressure at 100° is given by

$$p_{100} = H + h_{100},$$

and that at any unknown temperature t, by

$$p_t = H + h_t.$$

Now if the conditions assumed in the deduction of (125) and (126) had all been fully realized there would result very simply

$$\beta = \frac{P_{100} - P_0}{100\, P_0} = \frac{p_{100} - p_0}{100\, p_0} = \frac{h_{100} - h_0}{100\,(H + h_0)}; \qquad (133)$$

and

$$t = \frac{P_t - P_0}{\beta P_0} = \frac{p_t - p_0}{\beta p_0} = \frac{h_t - h_0}{\beta\,(H + h_0)}. \qquad (134)$$

But, in point of fact, p_{100}, p_0, and p_t are all slightly different from P_{100}, P_0, and P_t, for the latter correspond to an absolutely constant volume and to a condition in which all of the confined air undergoes the heating or cooling operation. Since neither of these conditions is realized in practice two corrections must be applied to the observed pressures. The first is the correction for the expansion of the bulb. If γ represent the cubical coefficient of expansion of glass, (i.e., the expansion per cc. per degree, a quantity which is equal to .000025), then the initial volume V of the bulb will have changed at 100° to $V + 100\gamma V\ [= V(1+100\gamma)]$. In order to reduce this volume back to V, it would be necessary to apply some pressure x such that (see Boyle's Law),

Correction for expansion of the bulb.

$$\frac{V(1+100\gamma)}{V} = \frac{x}{p_{100}}, \quad \text{or} \quad x = p_{100}(1+100\gamma). \qquad (135)$$

Applying this correction then to both (133) and (134) there results:

$$\beta = \frac{p_{100}(1+100\gamma) - p_0}{100\, p_0} = \frac{p_{100} - p_0}{100\, p_0} + \frac{\gamma p_{100}}{p_0}; \qquad (136)$$

and

$$t = \frac{p_t(1+\gamma t) - p_0}{\beta\, p_0} = \frac{p_t - p_0}{\beta\, p_0 - \gamma p_t}.* \qquad (137)$$

But for even moderately accurate results a still further correction must be applied to all of the observed pressures in order to make allowance for that portion of the confined air which escapes the changes in temperature which take place within the bulb. Thus it is evident that if the air in the capillary stem and in the space about c fell to the zero temperature, p_0 would be somewhat smaller than it is. Similarly if this air were heated with the rest to 100°, p_{100} would have a larger value. The error arising from this source is eliminated by multiplying the β and t given by (136) and (137) by a factor of the form

Correction for unheated portion of air.

$$1 + \frac{v}{V}\,\frac{p}{p_0}\,\frac{1}{1 + .00367t'},$$

in which v is the volume of the unheated portion of V, t' the temperature of the room near c, and p is p_{100} if the correction is applied to β, p_t if it is applied to t. The deduction of the form of this factor is comparatively easy but will scarcely be found profitable here (see Kohlrausch's Leitfaden der Praktischen Physik, 8th ed., p. 112).

Of course, if the barometer height is not 760 mm., the temperature of steam will not be exactly 100°, and a final modification must be made of (136) to take into account the change in boiling point with change in pressure. The correction will be made by replacing the 100 of equation

Correction for boiling temperature.

* The last expression is obtained by solving the preceding equation for t.

(136) by the value of the boiling point taken from table 6 in the Appendix.

The pressure coefficient β.

DIRECTIONS.—1. Lower C' (Fig. 77) until the mercury in the left arm of the manometer is below the three-way stopcock s, then open communication through r and n (see 77b), between B and the outside air. Attach to n a calcium chloride drying tube and an exhaust pump precisely as in the last experiment. When B has been thoroughly dried,* turn the cock so as to restore communication with the manometer. Then, with the aid of an ice scraper (see Fig. 78), prepare several quarts of pure shaved ice, pack it carefully about the bulb within any convenient vessel from which the water may be drained, *at the same time lowering C' so that the contraction of the air in B may not cause mercury to pass over into the bulb*, then pour over distilled water and repack. The water is added both to insure good contact and to make certain that the temperature of the ice is not below 0°. A mixture prepared in this way from shaved ice or clean snow gives a perfectly constant zero, but dry ice or snow can not be depended upon. Let the ice surround the whole bulb and the tube t up to the point to which the steam will have access when the bulb is immersed in the steambath. After a lapse of five minutes, adjust C' and e until the mercury just touches the platinum point c; then set the sliding index i successively upon the tops of the mercury menisci in the two arms, and take the corresponding vernier readings upon the vertical scale. Repeat both the setting and the observations, and see how closely successive values of h_0 can be made to agree. Next insert an asbestos screen between the bulb and manometer, replace the ice by the steam jacket shown in Fig. 77, and, as soon as the expansion has ceased, make again a series of settings and readings. In order to preclude the possibility of a leak, it is well, during the expansion, to keep the mercury always above the stopcock. After reading the barometer, obtain the volume of unheated air as follows: Place a small beaker underneath n and raise C' carefully until the mercury rises in the capillary tube t to the point to which this tube was

FIGURE 78

* Of course this drying need not be repeated with every experiment. The instructor will decide in each case whether it is necessary or not.

heated. Then turn the cock s and let mercury run out through n until the mercury level in the left arm has exactly reached the platinum point c. From the weight of the mercury which has emerged from s, v can at once be obtained. The volume of B, viz., V, can be calculated with sufficient accuracy from a careful estimate of the dimensions of the bulb. Or the bulb may be immersed in a graduate and the rise of the water noted. If v is of the order $\frac{V}{100}$ an error of as much as 10% in the measurement of V will introduce into the determination of β an error of but one-tenth of one per cent.

Correction of mercury thermometer.

2. Having found β, surround B with a water bath, immerse also in the bath the mercury thermometer which is to be standardized, seeing to it that the whole of the mercury in both bulb and stem is as nearly immersed as possible, and find the corrections of the mercury thermometer at about 25°C., 50°C. and 75°C. [see (137)]. In all of these determinations, *stir the water continually* and take no readings until a stationary condition has been reached. Next find the correction of the mercury thermometer at 0°, either by filling a large funnel with pure shaved ice, pouring over distilled water and immersing the thermometer up to its 0° mark; or better still by immersing it in a small vessel of distilled water (see Fig. 79) which is surrounded by a freezing mixture of ice and salt, and noting the point at which, with continual stirring, the mercury becomes stationary. In the latter case it will generally be found to fall below the freezing point and then to rise suddenly to it as the ice crystals begin to appear. Lastly find the correction at the boiling point by immersing the thermometer up to the mark 100 in a double-walled steam jacket like that shown in Fig. 77. The correction will be the difference between the observed boiling temperature and that given in the Appendix for the existing pressure. It will be positive if the mercury thermometer reads too low, negative if it reads too high. In all readings of thermometers, take great pains to avoid the error of parallax. This is done by placing the stem carefully at right angles to the line of sight.

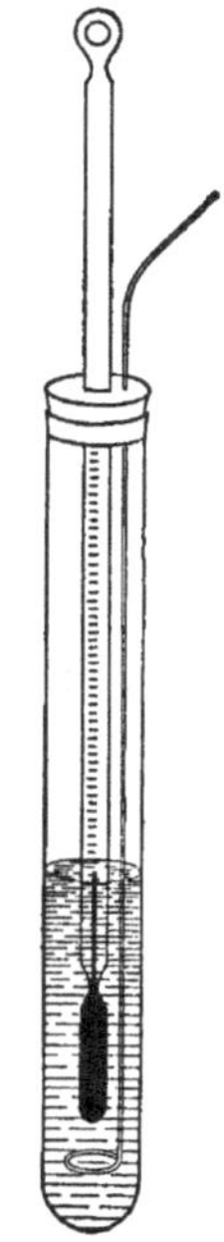
FIG. 79

From the five observed points plot a curve of thermometer corrections (cf. also the two-point curve of Figure 86, Ex. XVIII).

Record

1. Determination of β.

	Melting ice 1st setting	2d	Steam 1st setting	2d
Reading at c	= ——	——	——	——
Reading in rt. arm	= ——	——	——	——
$H =$ ——	$h_0 =$ ——	——	$h_{100} =$ ——	——

Mean $h_0 =$ —— $\therefore P_0 =$ —— Mean $h_{100} =$ —— $\therefore P_{100} =$ ——

$v =$ —— $V =$ —— $\therefore \beta =$ —— % error = ——

2. Correction of mercury thermometer. $H =$ ——

	0° 1st	2d	25° 1st	2d	50° 1st	2d	75° 1st	2d
At c	——	——	——	——	——	——	——	——
Rt. arm	——	——	——	——	——	——	——	——
$h_0 =$	——	——	$h_{t_1} =$ ——	——	$h_{t_2} =$ ——	——	$h_{t_3} =$ ——	——

Means h_0 —— h_{t_1} —— $\therefore t_1$ —— h_{t_2} —— $\therefore t_2$ —— h_{t_3} —— $\therefore t_3$ ——

By mercury thermometer t_1 —— t_2 —— t_3 ——

Correction of Hg. ther. at 0° — at 25° —— at 50° —— at 75° —— at 100° ——

Problems

1. From equation (126), which defines t, the measure of temperature upon the centigrade scale, show that, if the volume of a gas remain constant while the temperature changes,

$$\frac{P_{t_1}}{P_{t_2}} = \frac{273 + t_1}{273 + t_2};$$

or, if T denote the temperature measured from a point 273° below 0°C., i.e., from the absolute zero, show that

$$\frac{P_{t_1}}{P_{t_2}} = \frac{T_1}{T_2}. \tag{138}$$

2. Show also that if the pressure of a gas remain constant while its temperature changes, the volume is directly, the density inversely, proportional to the absolute temperature.

3. Since Boyle's Law gives the variation in density when the pressure alone changes, the temperature remaining constant, it is evident that from this law, and the rule deduced in the preceding problem, the density of a gas can be calculated for all temperatures and pressures if it has once been determined for one single value of the temperature and pressure. Thus, given that the density of air at 16°C., 745 mm., is .001192; find the density at 0°, 745 mm.; at 0°, 760 mm.; at 120°, 755 mm.

4. Find the volume occupied by 28.88 gm. of air at 0°, 760 mm.

5. How much work is done against atmospheric pressure when 10 gm. of air at 0°, 750 mm., are heated to 50°, 750 mm.?

6. In equation (132), nm is merely the mass per unit volume, i.e., the density of the gas. The density of air at 0°C., 76 cm., is .001293; hence, find from (132) the average velocity of the molecules of air at this temperature (p must, of course, be expressed in dynes).

7. Air is 14.44 times as heavy as hydrogen. Find the velocity of the hydrogen molecule. What relations do the densities of gases bear to their molecular velocities?

XVII

LAW OF AVOGADRO—DENSITIES OF GASES AND VAPORS

Theory

The speculations of the old Greek philosophers led some of them to the assumption of the atomic theory as to the constitution of matter; but it was not until 1803 that the English chemist, John Dalton, placed this theory upon an experimental rather than upon a purely speculative foundation. Experiments upon the combining powers by weight of different substances reveal four principles which find their most simple and natural interpretation in the atomic hypothesis. These are (1) the principle of constant proportions, (2) the principle of equivalence, (3) the principle of multiple proportions, (4) the principle of substitution.

Origin of the atomic hypothesis.

According to the first principle, *the proportions by weight in which the elements enter into a given compound are absolutely invariable.* For example, it is found that hydrogen will unite with chlorine, so as to leave no free hydrogen and no free chlorine, only when the number of grams of hydrogen present bears to the number of grams of chlorine one definite ratio. The same may be said of the combination of hydrogen with bromine or of hydrogen with iodine. These ratios are

The law of constant proportions.

$$\left.\begin{array}{l} 1 \text{ gm. hydrogen to } 35.18 \text{ gm. chlorine} \\ 1 \text{ gm. hydrogen to } 79.36 \text{ gm. bromine} \\ 1 \text{ gm. hydrogen to } 125.90 \text{ gm. iodine} \end{array}\right\} \cdot \qquad (138)$$

Similarly, the chlorides of potassium, sodium, and silver are always found to contain exactly the following proportions by weight:

$$\left.\begin{array}{l} 38.86 \text{ gm. potassium to } 35.18 \text{ gm. chlorine} \\ 22.88 \text{ gm. sodium to } 35.18 \text{ gm. chlorine} \\ 107.12 \text{ gm. silver to } 35.18 \text{ gm. chlorine} \end{array}\right\} \cdot \qquad (139)$$

It is evident that the interpretation of this law in the light of a molecular hypothesis as to the constitution of matter can only be that the atoms of each substance are of constant weight and that the molecules of compounds are always of the same atomic composition.

But another and still more significant relation is found to exist. From the examples given above, it is evident that 35.18 gm. of chlorine, 79.36 gm. of bromine, and 125.90 gm. of iodine may be called equivalent quantities in the sense that each one of them combines with exactly the same weight of hydrogen, viz. 1 gm. Similarly, 38.86 gm. of potassium, 22.88 gm. of sodium, and 107.12 gm. of silver may also be called equivalents, since each of them combines with the same quantity of chlorine. These latter substances can not be made to combine with hydrogen directly, but since the numbers given combine with just the number of grams of chlorine which has been found to be the combining equivalent of 1 gm. of hydrogen, these numbers may also be said to have been found in this indirect way to be the equivalents in combining power of 1 gm. of hydrogen. So much for the definition of *equivalent*. Now the fact of peculiar significance is this: A quantitative analysis of the *bromides* of potassium, sodium, and silver leads to precisely the same numbers for the equivalent weights of these substances as did a study of the *chlorides*. Thus the only proportions in which these substances will combine with bromine are:

The principle of equivalence.

$$\left.\begin{array}{ll} 38.86 \text{ gm. potassium to} & 79.36 \text{ gm. bromine} \\ 22.88 \text{ gm. sodium to} & 79.36 \text{ gm. bromine} \\ 107.12 \text{ gm. silver to} & 79.36 \text{ gm. bromine} \end{array}\right\} . \quad (140)$$

When, further, a study of the iodides leads again to the same three numbers, thus,

$$\left.\begin{array}{ll} 38.86 \text{ gm. potassium combines with} & 125.90 \text{ gm. iodine} \\ 22.88 \text{ gm. sodium combines with} & 125.90 \text{ gm. iodine} \\ 107.12 \text{ gm. silver combines with} & 125.90 \text{ gm. iodine} \end{array}\right\} , \quad (141)$$

it becomes certain that some very definite physical significance lies behind these numbers. The simplest possible interpretation to put upon them is, to take a particular case, that the particle of potassium which combines with chlorine to form the molecule of potassium chloride is exactly like the particle which combines with bromine to form the molecule of potassium bromide, and with iodine to form the molecule of potassium iodide.

If, now, it is decided to adopt a mere convention and call this particle an *atom* of potassium; if similarly the particle of bromine which enters into hydrobromic acid, and into the bromides of potassium, sodium, and silver be called an *atom* of bromine; and if a similar convention be adopted with reference to all of the substances thus far mentioned, then the weights of all these atoms in terms of the atom of hydrogen must be simply the numbers, above given, which represent the combining powers with reference to hydrogen, of the elements mentioned. Further, since it has been decided to regard this quantity of each element which enters into the molecule of any of the above compounds as a unit rather than as a combination of two or more units, the following symbols will henceforth be used: hydrochloric acid = HCl, hydrobromic acid = HBr, hydroiodic acid = HI, potassium chloride = KCl, sodium chloride = NaCl, silver chloride = AgCl. Corresponding formulae for the bromides and iodides are: KBr, NaBr, AgBr, KI, NaI, AgI. Of course, these particles which enter into the above combinations may themselves be aggregations of 2, 3, 10, or 1000 smaller particles for aught we know, but so long as no evidence is brought forward to show that some sort of compound substance exists, the molecule of which contains a smaller amount of chlorine, for example, than that quantity which is found in HCl, KCl, NaCl, and AgCl, this quantity will be called by common consent, i.e., by definition, the atom of chlorine. An atom would then be defined as the smallest particle of any element which is known to enter into the molecule of any compound.

The facts of equivalence which have been above presented, and which constitute one of the strongest arguments for the atomic hypothesis, may be summarized thus: *The study of many different compounds leads often to precisely the same number as the combining equivalent of a given element with reference to hydrogen.*

But in some cases the study of different compounds leads to *different* numbers for the equivalent of a given element with reference to hydrogen. For example, Dalton found that olefiant gas yielded upon decomposition the two elements carbon and hydrogen in the proportions by weight, 6 carbon to 1 hydrogen, while marsh gas yielded the same

Law of multiple proportions.

two elements in the ratio 3 carbon to 1 hydrogen. Further study of other compounds revealed the fact that *whenever elements have the power of combining in different proportions, these proportions always bear simple ratios to one another.* This is known as the *law of multiple proportions.* The self-evident interpretation of the law, in the light of a molecular hypothesis, is that it is possible in some cases for two, or three, or some other *small* number of atoms of a given element to enter into the constitution of a compound molecule. It was probably the discovery of this law of multiple proportions which first convinced Dalton of the truth of the atomic theory. To illustrate it by a further example, there are four different compounds of the elements chlorine, potassium, and oxygen in which the proportions by weight of the three elements are as follows:

CHLORINE	POTASSIUM	OXYGEN	
35.18	38.86	15.88	
35.18	38.86	31.76	(142)
35.18	38.86	47.64	
35.18	38.86	63.52	

It will be observed that in all four compounds the potassium and chlorine have exactly the same ratios as in KCl. The simplest and most natural interpretation is that all of the compounds contain an atom of potassium and an atom of chlorine. The smallest amount of oxygen found in the molecule of any of these compounds weighs then 15.88 times as much as the hydrogen atom. If this amount be called the oxygen atom (at least until some smaller amount is found to enter into some other sort of molecule), then the second compound contains two atoms of oxygen, the third three, the fourth four, and the formulae for the four substances are $KClO$, $KClO_2$, $KClO_3$, and $KClO_4$.

The principle of substitution. It will have been already observed that the fact of combination in multiple proportions introduces an uncertainty into the determination of the true combining equivalents, i.e., the atomic weights of some elements. For example, from Dalton's experiments on olefiant gas and marsh gas, it might be inferred that the atomic weight of carbon was 6 and the formula for olefiant gas CH and for marsh gas CH_2. But 3 might with equal reason be taken as the atomic weight of carbon, and C_2H

and CH as the corresponding formulae. The following experiment, however, makes it certain that the molecule of marsh gas contains at least four hydrogen atoms. It is found that, by successive treatments with chlorine, marsh gas can be made to yield five different compounds in which hydrogen, carbon, and chlorine are combined in the following proportions by weight:

HYDROGEN	CHLORINE	CARBON
4	0	11.91
3	35.18×1	11.91
2	35.18×2	11.91
1	35.18×3	11.91
0	35.18×4	11.91

(143)

The existence of this series proves that the hydrogen in marsh gas is divisible into at least four parts, and that 1, 2, 3, or 4 of these parts may at will be replaced by 1, 2, 3, or 4 atoms of chlorine. It is certain, then, that there must be as many as four atoms of hydrogen in the molecule of marsh gas. If the quantity of carbon present in this molecule be provisionally assumed to be the atom (an assumption which is justified by the study of the other carbon compounds), the formula for marsh gas becomes CH_4; for olefiant gas, CH_2; and the hydrogen equivalent of carbon, i.e., its atomic weight, is fixed not at 6 or 3 but at 12 (accurately at 11.91).

Again, since water is found to contain hydrogen and oxygen in the ratio 1 to 7.94, this might at first be taken as the ratio of the weights of the atoms of hydrogen and oxygen, and the symbol for water written HO. This would indeed be inconsistent with the interpretation put upon the series (142), from which it was inferred that the atomic weight of oxygen was 15.88. It would be possible, however, to reconcile this difficulty by assuming the formulae for the compounds in (142) to be $KClO_2$, $KClO_4$, $KClO_6$, $KClO_8$,—an assumption not very natural, it is true, since the new formulae at once suggest that it is O_2 rather than O, which is the oxygen unit. But the principle of substitution leaves no doubt as to which of the above possibilities must be chosen. For if potassium be allowed to act on water, a portion of the hydrogen is drawn off and replaced by potassium; if it be allowed to act again, all of the hydrogen is replaced, the proportions by weight in the three compounds being as follows:

HYDROGEN	POTASSIUM	OXYGEN	
2	0	15.88	
1	38.86	15.88	(144)
0	77.72	15.88	

This series makes it certain that there are at least two atoms of hydrogen in the water molecule, and therefore supports the first conclusion that the atomic weight of oxygen is 15.88, and the constitution of water H_2O. *This gradual replacement of a given element of a compound by successive reactions is called substitution.*

Some other atomic weights and chemical formulae.

After the atomic weights of carbon and oxygen have been fixed at 11.91 and 15.88, the discovery that carbon monoxide contains these elements in exactly these proportions, and that carbon dioxide contains the same elements in the proportions 11.91 to 31.76, leads at once to the formulae CO and CO_2 to represent the chemical constituents of these gases. Again, when carbon and nitrogen are found to unite in the ratio 11.91 to 13.93, and carbon, nitrogen, and hydrogen in the ratio

CARBON		NITROGEN		HYDROGEN
11.91	:	13.93	:	1

it is natural to take 13.93 as the probable value of the weight of the nitrogen atom. Then the formula for nitric oxide, which contains 13.93 gm. of nitrogen to 15.88 gm. of oxygen, becomes NO, and that for nitrous oxide, in which nitrogen and oxygen are found in the ratio 13.93 to 7.94, becomes N_2O. Furthermore, the fact that sulphur combines with chlorine in the ratio 31.83 to 35.18, and also with bromine in the ratio 31.83 to 79.36, suggests 31.83 as the atomic weight of sulphur.

The law of Avogadro.

Enough has now been said to show how, aided only by the laws of combination, chemists, beginning with Dalton, set about devising tables of atomic weights and molecular constitutions; tables which, to be sure, were only provisional, since it might sometimes be difficult to determine whether the atomic weight of a substance should be represented by a certain number, or by some simple multiple or sub-multiple of that number. But as more compounds were investigated, the choices became more and more restricted, and it can scarcely be doubted that the study of combining powers alone would have led ultimately to most of the now accepted values of atomic weights and chemical formulae, even if Avogadro's Law had never been discovered. The discovery of this law, however, facilitated greatly the work of fixing these quantities. The law was announced in 1811 by the Italian chemist whose name it bears. *It asserts that at a given temperature and pressure all gases contain the same number of molecules per cubic centimeter.* The proof of the law rests upon a remarkable relation which is found to exist between the combining powers of substances and their gas or vapor densities. The following table* brings out clearly this striking relation. The column headed "Density" represents the results of experiments upon the relative densities, at a given temperature and pressure, of a number of the gases already mentioned. For convenience of representation, the density of hydrogen gas is taken as 2.

GAS	DENSITY	ATOMIC WEIGHT
Hydrogen	2	1
Nitrogen	27.82	13.93
Oxygen	31.80	15.88
Chlorine	70.72	35.18
Bromine	159.54	79.36
Iodine	254.73	125.90
Sulphur	64.06	31.83
		MOLECULAR WEIGHT
Hydrochloric acid (HCl)	36.30	36.18 (1 + 35.18)
Marsh gas (CH_4)	16.08	15.91 (11.91 + 4)
Carbon monoxide (CO)	27.95	27.79 (11.91 + 15.88)
Carbon dioxide (CO_2)	44.10	43.67 (11.91 + 15.88 × 2)
Nitric oxide (NO)	29.95	29.81 (13.93 + 15.88)
Nitrous oxide (N_2O)	44.10	43.74 (13.93 × 2 + 15.88)
Water (H_2O)	18.03	17.88 (2 + 15.88)

It is seen that in the lower group of substances the numbers which represent vapor densities in terms of a gas one-half as dense as hydrogen are throughout almost identical with the numbers which represent the weights of the molecule in terms of the weight of the hydrogen atom, as above defined. But if the weights of the individual molecules of a number of gases bear the same ratios

*See Landolt and Börnstein, *Physikalisch-chemische Tabellen*, pp. 115, 116; and Wüllner, *Experimental Physik*, Vol. II, p. 802.

as the weights of the gases per cc., then evidently the number of molecules per cc. must be the same in all the gases. This remarkably simple conclusion, which applies necessarily to all of the gases of the *second* group, provided the conclusions above reached as to their molecular constituents are correct, is seen to apply also to all the gases of the first group, if only the molecules of the gases, hydrogen, nitrogen, oxygen, chlorine, bromine, iodine, and sulphur, be assumed to be twice as heavy as the atoms of these substances; that is, if these molecules are composed each of two atoms, thus, H_2, N_2, O_2, Cl_2, Br_2, I_2, S_2; for then the molecular weights become 2, 27.86, 31.76, 70.36, 158.72, 251.80, and 63.66 respectively, numbers which are in remarkably close agreement with those given in the column of densities.

Now, it is found that with equally simple choices as to the molecular constitutions of those gases in which the combining powers of the constituent elements leave two or more choices open, *the densities of all known gases become identical with their molecular weights.* This constitutes overwhelming evidence for the truth of Avogadro's hypothesis. *It is this fact of agreement between molecular weights and vapor densities which is the experimental basis for the law of Avogadro.** This agreement is least perfect in the cases of those gases which show the largest departures from Boyle's Law. For *actual* gases this law, therefore, like those of Boyle and Gay-Lussac, is only a close approximation.

Experiment

Object. 1. To determine the density of CO_2 and to compare the same with its molecular weight.

2. To determine the density of water vapor and to compare the same with its molecular weight.

A glass globe of known volume V is weighed, first when full of air at temperature T_1 (absolute), pressure P_1, then when full

Method. of the unknown gas at temperature T_2, pressure P_2. In these weighings a closed bulb of the same volume as

*The proof of this law which Maxwell first drew from the kinetic theory—a proof which rests upon the Maxwell-Boltzman law of the distribution of energies between two sets of unlike particles in a gaseous mixture, and which has since been given a place in a large number of chemical and physical texts—can not be recognized as adequate. (See Note by Rayleigh in Maxwell's *Theory of Heat*, 10th ed., p. 326.)

the density globe is used as a counterpoise so as to eliminate all effects due to changes in the buoyancy of the air. If the difference between the first and second weighings be represented by w (w being of course negative if the second weight exceeds the first), the density of the gas at T_2P_2 by d_{g_2}, the density of air at T_1P_1 by d_{a_1}, then evidently

$$Vd_{a_1} - Vd_{g_2} = w.* \tag{145}$$

Further, since density is directly proportional to pressure when the temperature remains constant (see Boyle's Law), and inversely proportional to absolute temperature when the pressure remains constant (see Ex. XVI, Problem 2), it is evident that the equation which expresses the relation between the densities of air at T_1P_1 and at T_2P_2 is

$$\frac{d_{a_1}}{d_{a_2}} = \frac{P_1}{P_2}\frac{T_2}{T_1} \tag{146}$$

Now, the quantity which will be first sought in this experiment is the density of the unknown gas in terms of the density of air at the same temperature and pressure, i.e., $\frac{d_{g_2}}{d_{a_2}}$. This is obtained easily from (145) and (146). Thus, substitution in (145) of the value of d_{a_1} obtained from (146) gives

$$Vd_{a_2}\frac{P_1}{P_2}\frac{T_2}{T_1} - Vd_{g_2} = w.$$

Whence

$$\frac{d_{g_2}}{d_{a_2}} = \frac{P_1}{P_2}\frac{T_2}{T_1} - \frac{w}{Vd_{a_2}}. \tag{147}$$

All of the quantities on the right side of (147), excepting d_{a_2}, are measured directly in the experiment. d_{a_2} is obtained from (146) and the result of Ex. XV; or, if it is found more convenient to determine d_{a_1} than d_{a_2}, (147) may, with the aid of (146), be thrown into the form

$$\frac{d_{g_2}}{d_{a_2}} = \left(1 - \frac{w}{Vd_{a_1}}\right)\frac{P_1}{P_2}\frac{T_2}{T_1}. \tag{148}$$

*In this equation the expansion of the bulb is neglected, because in neither of the following determinations will it affect the result by more than a small fraction of one per cent. If it is desired to take it into account it is only necessary to replace (145) by $Vd_{a_1} - V(1+\gamma t)d_{g_2} = w$ (in which γ is the expansion coefficient of glass and t the number of degrees between T_1 and T_2), and then to solve precisely as above.

Finally, since air is 14.44 times heavier than hydrogen, it is only necessary to multiply the density of the unknown gas in terms of air by 28.88 in order to obtain its density in terms of a gas one-half as heavy as hydrogen. This is the quantity which, according to the law of Avogadro, should agree with the molecular weight.

DIRECTIONS.—1. For convenience in filling and weighing, the density globe, the capacity of which is about 250 cc., is provided with two taps *a* and *b* (see Fig. 80). The volume of this globe is first to be found by filling it with water and weighing upon the trip scales. The density of water at the observed temperature is to be taken from the table of water

To find V.

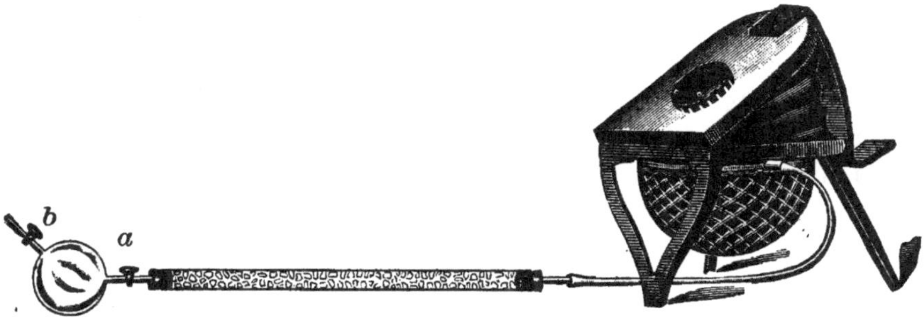

FIGURE 80

densities in the Appendix. *V* is, of course, the ratio of the weight and density of the water.

Having found *V*, dry the globe by carefully rinsing it with alcohol and then forcing through it a current of air from a bellows, at the same time heating gently by means of a rapidly moving Bunsen flame. Continue this operation until all odor of alcohol has disappeared from the bulb. Then, after carefully lubricating the stopcocks, connect the bulb with the bellows through a calcium chloride drying-tube, as in Fig. 80, and force through the combination a *very gentle* current of air for about one minute. Then close tap *b* and allow the apparatus to stand in this condition for about five minutes, shielding the bulb as much as possible from temperature changes. Next close tap *a*, read the barometer, and take the temperature by means of a thermometer hung near the bulb. Then detach the bulb, carefully remove all dust and grease from

Filling with air and weighing.

its surface, and weigh upon an analytical balance, using a counterpoise of the same volume as the globe and following the directions given in Ex. XV for the first weighing. It is particularly important that no mercury be allowed to touch the bulb, as it is nearly impossible to remove small mercury globules from a glass surface.

After weighing, put the bulb into connection, through the drying-tube, with a reservoir of CO_2 or with a vessel in which the gas is being generated; and, keeping the bulb in such position that the exit tap is on top, allow a gentle current of the gas to flow through the bulb for about two minutes. Find the temperature of the issuing gas by placing, at the orifice of the exit tap, the bulb of the thermometer previously used. Then close first the entrance, then the exit tap; detach the bulb and re-weigh, following the directions given in Ex. XV for the second weighing. If neither the temperature nor the pressure differs appreciably from the values taken in connection with the first weighing, (147) reduces to

Filling with CO_2 and weighing

$$\frac{d_{g_2}}{d_{a_2}} = 1 - \frac{w}{Vd_{a_2}},$$

w being itself negative in this case, since CO_2 is heavier than air.*

2. The method here used for determining the density of water vapor in terms of air does not differ in principle from that employed with CO_2. Since, however, the maximum pressure which water vapor is able to exert at ordinary temperatures is less than atmospheric pressure (see Ex. XVIII), it is evident that it must be impossible under ordinary atmospheric conditions wholly to replace the air in the density globe by water vapor. This can be done easily at any temperature at which the maximum pressure of water vapor is more than atmospheric, e.g., at 150°C. Hence the following directions: After carefully drying the bulb B (see Fig. 81), by repeatedly

Filling the bulb with water vapor and weighing.

*Precisely the same method may be used with gases lighter than air, save that in this case the exit tap should be at the bottom during the filling. However, all of the standard determinations of gas densities have been made with bulbs provided with but one tap rather than with two as here described. The bulb has then been completely exhausted before being put into connection with the gas reservoir.

warming and exhausting through a calcium chloride tube as in Ex. XV, make a first weighing upon an analytical balance, using, as above, a counterpoise of the same volume as *B*. Then place the capillary orifice *o* beneath the surface of distilled water, and warm the bulb slightly by means of the hand. Upon cooling, one or two cubic centimeters of water will be drawn into the bulb. Next, completely immerse the bulb in melted paraffin (see Fig. 81), leaving but a few centimeters of the capillary tube projecting above the surface of the liquid. Keep the temperature of the bath constant, at, say, 120° C., by very thorough and continuous stirring, and by a proper regulation of the Bunsen burner. The rapid vaporization of the water will drive all air from *B*. If a flame be held in front of *o*, it will be seen to be deflected by the rapid current of issuing steam. When the cessation of this deflection indicates that the water in *B* has entirely boiled away, seal the globe by means of a fine blow-pipe flame. This is best done by heating the tube to softness just above the paraffin and then drawing off the tip. As soon as the sealing is complete, take readings of the temperature of the bath and of the barometer height. Then remove the bulb, clean it thoroughly with a cloth while the paraffin is still hot, and test for a leak by allowing the condensed steam to run down to the tip of the tube and observing whether or not fine bubbles enter the bulb. Then, after cooling to the temperature of the room, again weigh the bulb together with the drawn-off tip.

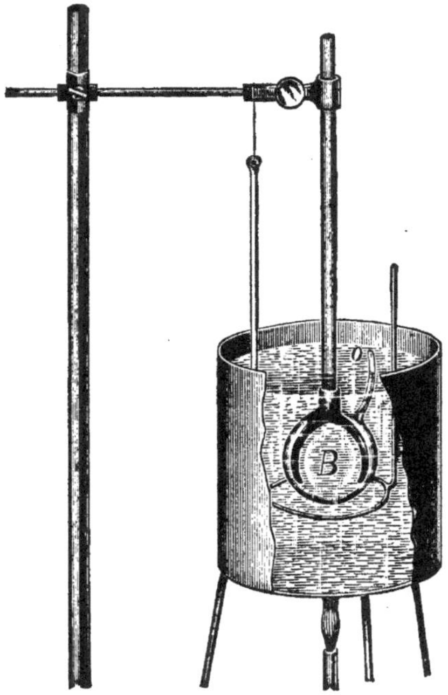

FIGURE 81

To find the volume of the bulb, file off under water the sealed tip. The bulb will at once fill with water. Weigh this bulb full of water upon the trip scales, and compute the volume as in the experiment with CO_2. If the bulb does not completely fill with water, the filling may be completed by means of a pipette. It is true that (147) and (148) are not then rigorously correct, but unless the bubble is quite large, the error introduced will be negligible.

Volume of bulb.

Record

1. Weight on trip scales of bulb = ——— of bulb + water = ———
Temperature of water = —— ∴ density of water = —— ∴ V = ——
P_1 = ——— T_1 = ——— P_2 = ——— T_2 = ———
Wt. added to counterpoise to balance globe + air = —— rest'g pt. = ——
Wt. added to counterpoise to balance globe + CO_2 = —— rest'g pt. = ——
Rest'g pt. after addition of 2 mg. = —— ∴ sensitiveness = —— ∴ w = ——
∴ $\frac{dg_2}{da_2}$ = ——— × 28.88 = ——— molecular wt. = ——— % error = ———

2. P_1 = ——— T_1 = ——— P_2 = ——— T_2 = ———
Wt. added to counterpoise to balance globe + air = —— rest'g pt. = ——
Wt. added to counterpoise to balance globe + H_2O = —— rest'g pt. = ——
Rest'g pt. after addition of 2 mg. = —— ∴ sensitiveness = —— ∴ w = ——
Weight on trip scales of bulb = ——— of bulb + water = ———
Temperature of water = —— ∴ density of water = —— ∴ V = ———
∴ $\frac{dg_2}{da_2}$ = ——— × 28.88 = ——— molecular wt. = ——— % error = ———

Problems

1. One gram of air was introduced into an empty spherical bulb of radius 10 cm. Find the pressure, in mm. of mercury, which the gas exerted against the walls of the bulb when the temperature was 25°C.

2. Find the pressure which the same weight of N_2O gas would exert in the same vessel at the same temperature. Compute similarly for hydrogen gas; for CH_4 gas.

3. One gram of nitrogen and 1 gram of H_2S are introduced together into the bulb used in the first Problem. Find the pressure at 25°C

4. Find the volume which 2 grams of hydrogen will occupy at 0°C. 76 cm. The same for 32 grams of oxygen; for 30 grams NO. Explain the connection between the results.

5. Find the density of air at 100°C. 76 cm.; of water vapor. If the density of water at 100°C. 76 cm. is .95852, find how many times water expands upon vaporizing.

6. Find the density of alcohol vapor at 72° 76 cm., the formula for alcohol being C_2H_6O. If the density of alcohol at 72° is .8, find how many times alcohol expands upon vaporizing.

XVIII

THE PRESSURE-TEMPERATURE CURVE OF A SATURATED VAPOR

Theory

The kinetic theory of liquids.

If the molecules of gases and of vapors are in rapid motion, the molecules of liquids must be also, for no fundamental distinction exists between the liquid and the gaseous conditions. At high temperatures the two states become absolutely identical. At temperatures below a certain critical value, however, the possession of a clearly marked surface may be taken as the distinguishing feature of a liquid.

Figures 82 and 83 illustrate the probable differences between the motions of the molecules in gases and in liquids at ordinary temperatures and pressures. In the former (Fig. 82) the molecules are so far apart that their mutual attractions may in general be neglected. They move in straight lines through distances which are large in comparison with their own dimensions. Their motions change direction at collision only. The zigzag line represents a possible path of one single molecule in going from position 1 to position 1′ and making impacts in so doing against molecules 2, 3, 4, etc., up to 12. The mean distances traversed between

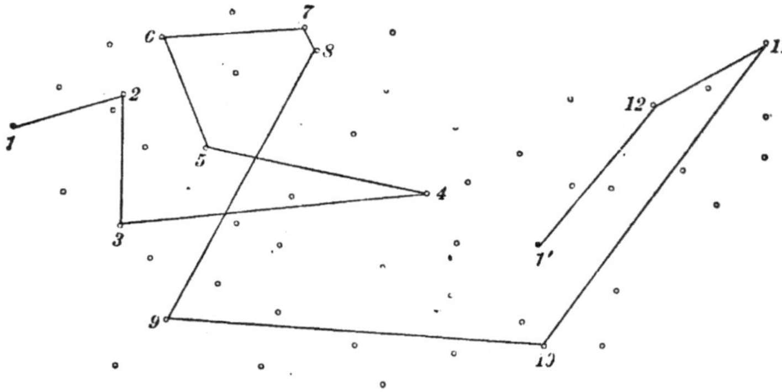

FIGURE 82

successive collisions by a molecule of air at 0°C., 76 cm., is only about .00006 mm., but this is a distance which, small as it is, is at least 100 times as large as the diameter of the molecule. In liquids, on the other hand (see Fig. 83), the molecules are crowded so closely together that their motions between impacts are extremely minute—of the same order of magnitude as the molecules themselves—and at the surface of the liquid, where there is greater freedom of motion, the paths of the particles are influenced not only by collisions but also by the attractions of the other molecules. On account of the enormous number of molecules present in or near the surface, this downward force upon a molecule just above the surface is doubtless very large; so large, in fact, that the molecules which are moving away from the surface are in general unable to leave it. They simply rise to a certain distance by virtue of their velocities, after the manner of projectiles shot up from the earth, and then fall back again into the liquid. Their paths near the surface thus become similar to the forms shown in Fig. 83. The zigzag line in the figure represents a possible path of a particle in the body of a liquid.

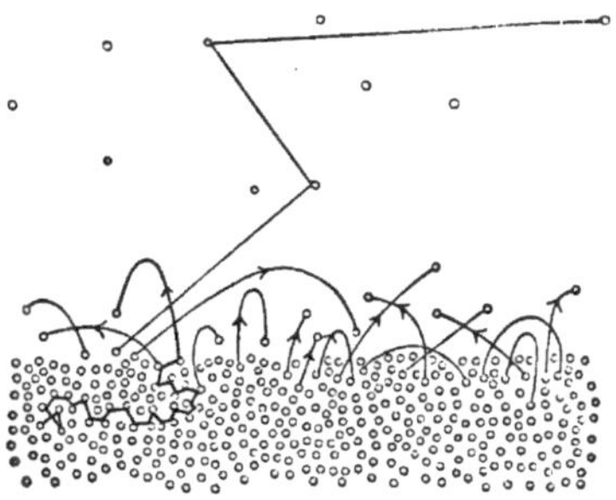

FIGURE 83

The kinetic theory and vaporization.

But conditions may arise which render possible the escape of a molecule from the liquid; for it must not be assumed, either in the case of liquids or of gases, that the molecules of the same substance all have the same velocity, for even if they were all given a common velocity to begin with, the collisions would at once create differences. Again, although constancy of temperature means that the *mean* velocity remains the same, the velocity of each individual molecule will in general change at each impact. The conditions of impact must often be such that a molecule receives a velocity much greater than the mean value. If the substance be a liquid, and if one of these more rapidly moving molecules be near the surface, it may be able to break through the thin space in which the powerful attraction exists, and to move off as an independent molecule into the space above. It thus happens that the space above the liquid in a closed vessel gradually becomes filled with the gaseous form of the substance of the liquid. This gas becomes more and more dense as more molecules escape, but there is evidently a limit to its possible density, for many of the escaped molecules chance, in their wanderings, to return to the surface and reënter the liquid. The number of molecules thus returning per second evidently increases as the number of molecules above the liquid increases, i.e., it is proportional to the density of the vapor. When this density has reached a certain limit, there is set up a condition of "active equilibrium" in which as many molecules reënter the liquid per second as escape. When this condition is reached, the vapor is said to be *saturated*, that is, it has the largest density which it is ever able to have at the existing temperature, and it therefore exerts the largest pressure which it ever can exert at this temperature.

Dependence of the density and pressure of a saturated vapor upon temperature.

But, if the temperature be raised, the vapor can evidently have a larger density, for the number of molecules escaping per second must be greater at the higher temperature, i.e., at the higher mean velocity, and hence the density of the vapor must be greater before the condition of equilibrium is set up. Also, since the pressure exerted by the vapor is proportional both to the density and to the mean kinetic energy of each impact, and since both density and kinetic energy increase with temperature, it is evident that the pressure must rise with two-fold rapidity as the temperature rises.

Density and pressure independent of volume.

But, if the temperature be held constant, all attempts to increase the density or the pressure of a vapor which is in contact with its liquid in a closed vessel must be futile. To see this clearly, conceive of a few drops of ether inserted into a barometer so as to fill the space above the mercury with ether and saturated ether vapor (see Fig. 84). As soon as the density of this vapor is temporarily increased by compression, i.e., by lowering the tube in the cistern, the equilibrium at the surface is destroyed, and immediately more molecules begin to enter the liquid per second than escape from it. Hence, in a very short time, enough ether condenses to restore the old condition of density and pressure. Similarly, raising the tube and thus increasing the volume occupied by the vapor diminishes the number of molecules which reënter the liquid per second, while leaving the number which escape, unchanged; hence equilibrium is soon reëstablished at the old density and pressure. This can be proved easily by observing that raising or lowering the tube does not alter the height of the mercury in the tube above the mercury in the cistern.

FIGURE 84

Retarding influence of air.

This readjustment to equilibrium conditions takes place almost instantly when the space contains only the liquid and its vapor. The presence of air or of any other gas exerts a very marked influence upon the time required for adjustment, but does not affect the ultimate result. Thus, in Fig. 84, the density and pressure of the ether vapor at a given temperature are ultimately the same whether air is present in the tube or not, i.e., just as much liquid will evaporate into a space filled with air as into a vacuum. But whereas, when the ether is introduced into a vacuum, the maximum density of the vapor is reached in a few seconds, when it is introduced into a space containing air, the condition of saturation may not be reached for a number of hours. Of course, if air be present, the total pressure against the walls is the sum of the pressures of the air and of the vapor.

Effect of air currents upon evaporation.

If the liquid be contained in a vessel which is open to the air, so that the pressure can not rise above the atmospheric condition, the vapor which forms is continually being removed by diffusion and by air currents, so that it never reaches its maximum density, i.e., the rate of exit of molecules from the liquid always remains greater than the rate of entry. Hence the liquid gradually disappears or "evaporates." The rate of this evaporation evidently depends upon the rapidity with which the vapor above the liquid is removed. Hence it is that fanning greatly facilitates evaporation.

The cooling effect of evaporation.

The cooling effect of evaporation is very easily understood in the light of the kinetic theory. For, since it is only the more rapidly moving molecules which are able to break away from the attraction which exists near the surface, the mean kinetic energy of the molecules of the liquid is continually being diminished by the loss of the most energetic members. And since temperature is a function of the average molecular energy, this loss means, of course, a continual reduction of the temperature of the liquid. This fall of temperature would continue indefinitely if the liquid did not soon become so much cooler than the surrounding objects that it receives heat from them as rapidly as it loses it by evaporation.

This passage from the liquid to the vaporous condition by surface evaporation takes place to some extent at all temperatures whenever the space above the liquid is not saturated. As the temperature is increased, outside conditions remaining the same, it takes place more and more rapidly, until finally a temperature is reached at which it begins to take place not simply at the surface but also within the body of the liquid, i.e., bubbles of vapor begin to form *beneath* the surface upon the sides of the containing vessel, whence they rise to the top, growing rapidly as they ascend. It is at once evident that this condition can not be reached until the maximum pressure exerted by the vapor which is formed from the liquid is at least equal to the outside pressure; for, if the pressure exerted by the vapor in the bubbles were less than that outside, these bubbles, even if formed, would at once collapse. *This temperature, then, at which the pressure of the saturated vapor becomes equal to the outside pressure, is called the boiling temperature.* It does not follow, however, that a liquid will always boil as soon as its temperature reaches the boiling point as here defined. In fact, the temperature of a boiling liquid must always be at least a trifle higher than that at which the pressure of the saturated vapor is equal to the outside pressure, for the pressure within the bubble must be sufficient to overcome not only the outside pressure but also the weight of the superimposed liquid and the surface tension of the bubble (see Ex. XXI). When the bubble, however, rises to the surface and breaks, the pressure exerted by the vapor which was contained within it must fall exactly to the atmospheric condition, and the temperature of this vapor must also fall, by virtue of expansion, to that temperature at which the pressure of the saturated vapor is equal to the existing atmospheric pressure. Hence, *a thermometer which is to indicate the true boiling temperature according to the above definition must be placed not in the boiling liquid itself but in the vapor rising from it.*

The boiling temperature.

The temperature of the liquid itself is a very uncertain quantity. Gay-Lussac found that the temperature of boiling water in a glass vessel was usually 1° to 3° higher than in a metal vessel. For the reasons above mentioned, it must always be a trifle higher than the boiling point; but under some circumstances it may rise many degrees above this temperature. For it is by no means necessary that bubbles of vapor begin to form as soon as the temperature is reached at which they are able to exist after being formed. The presence of air in the water or occluded in the walls of the containing vessel is found to be essential to the genesis of bubbles. A Frenchman named Donny found, in 1846, that when he very carefully removed this air he could raise the temperature of water in a glass vessel to 138°C. before boiling began. But in all such cases, since the pressure of the saturated vapor corresponding to the temperature of the water is much more than the atmospheric pressure, as soon as a bubble once starts it grows with explosive rapidity and produces the familiar phenomenon of "boiling with bumping." In 1861 another Frenchman, Dufour, succeeded in raising globules of water immersed in oil to a temperature of 175°C.

The temperature of the liquid.

Experiment

Object. To determine the variation of the boiling point with the pressure, (1) by the static method; (2) by the dynamic method.

DIRECTIONS.—1. The apparatus used in the static method is shown in Fig. 85. The bulb *B*, originally open at *c*, was first half filled with mercury. The long arm (about 5 mm. in diameter), also originally open at the top, was then exhausted and inclined till the mercury completely filled it up to a point at which it had been drawn down to capillary dimensions. The tube was then sealed off at this point, so that when the instrument was vertical the difference between the levels of the mercury in the bulb and in the tube was equal to the barometric height. Water was then inserted at *c* and boiled until the air was all driven from the bulb, when the opening at *c* was sealed off. Since, then, only water and water vapor exist above the mercury in the bulb, the difference between the levels in the

Filling the bulb.

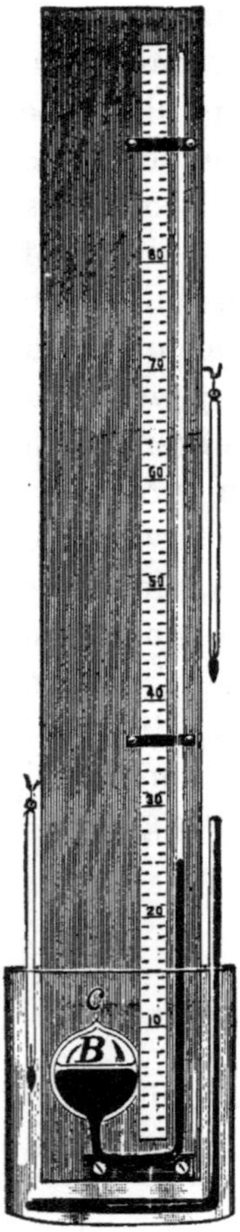

FIGURE 85

bulb and in the tube gives at once the pressure of the saturated water vapor in the bulb. It is, therefore, only necessary to vary the temperature of the bulb in order to obtain the curve expressing the relation between the temperature and the pressure of saturated water vapor.

The readings.

The whole bulb is to be placed in a jar of water whose temperature is first to be lowered, by the insertion of ice, to nearly 0°C., and then slowly raised to about 50°C. by pouring in hot water and siphoning off the cold. At about 50° it is well to replace the glass jar by a metal pail and thenceforth to heat slowly to 100° by means of a Bunsen flame. Between 0° and about 70°C. readings of the pressure are to be taken at intervals of 10° to 12°, between 70° and 100° at intervals of about 4°. The water must be very vigorously stirred throughout the experiment, and the temperature should be held constant for at least one minute before a reading is taken.

Corrections for errors in the thermometer.

The thermometer readings must all be corrected (1) for the errors of the instrument itself, and (2) for the length of the exposed thread of mercury. If the first correction is to be made accurately, the thermometer should be one which has been compared with a standard, as in Ex. XVI. If, however, this comparison has not been made, the corrections may be obtained with a moderate degree of accuracy by observing the corrections at the freezing and boiling points and interpolating between these points for the corrections at other temperatures. Thus, suppose that when wrapped in one layer of flannel and packed in shaved ice over which a little water has been poured, the thermometer reads –0.2°, and that when completely immersed to the top of the thread in a steam bath upon a day on which the boiling point of water should be 99.4°, the reading is 100.5°. The corrections at 0° and 100° are then +.2 and –1.1 respectively. If these two corrections be plotted as ordinates upon a horizontal line representing temperatures, and if the ends of these ordinates be connected by a straight line, as in Fig. 86, the corrections for intermediate temperatures may be read off

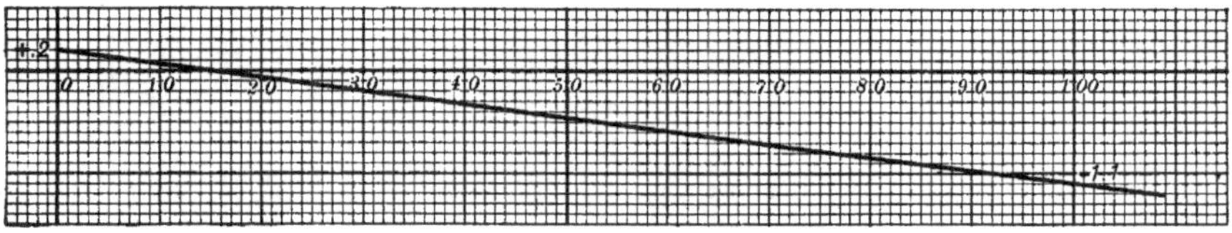

FIGURE 86

from this curve. Thus, in this case, the correction at 10° is +.1, at 40° it is –.3, at 95° it is –1.05, etc.

Correction for the exposed thread.

The correction for the exposed thread is obtained by adding to the observed temperature $.00016l(t - t_0)$, in which l is the length in degrees of the exposed thread of mercury, t the observed temperature corrected according to (1), t_0 the mean temperature of the exposed column obtained from a second thermometer whose bulb hangs about the middle of this column, and .00016 the apparent expansion coefficient of mercury in glass [.000181 (= coef. of Hg) – .000025 (= coef. of glass) = .00016, approximately].

Reduction of the observed pressures to 0°C.

In order to compare the results with tabulated values of vapor pressures, it is necessary to express all pressures in terms of columns of mercury at 0°C. This correction is made as usual by multiplying the observed heights by the ratio of the densities of mercury at the mean temperature of the observed column and at 0° (see Appendix). This correction may be made very roughly, for it is only at the higher temperatures that it amounts to more than the observational errors. The mean temperature of the column may be taken from a third thermometer hung near its middle point.

Correction for capillary depression.

The observed pressure will need a still further correction on account of the capillary depression of the mercury in the tube. This correction may be taken from the table in the Appendix.

2. The dynamic method consists in the direct observation of the temperatures of the steam which rises above a boiling liquid made to boil under varying pressures. In Fig. 87, A represents an air-tight metal boiler, which may be replaced if need be by a simple long-necked glass flask. B is a condenser through which

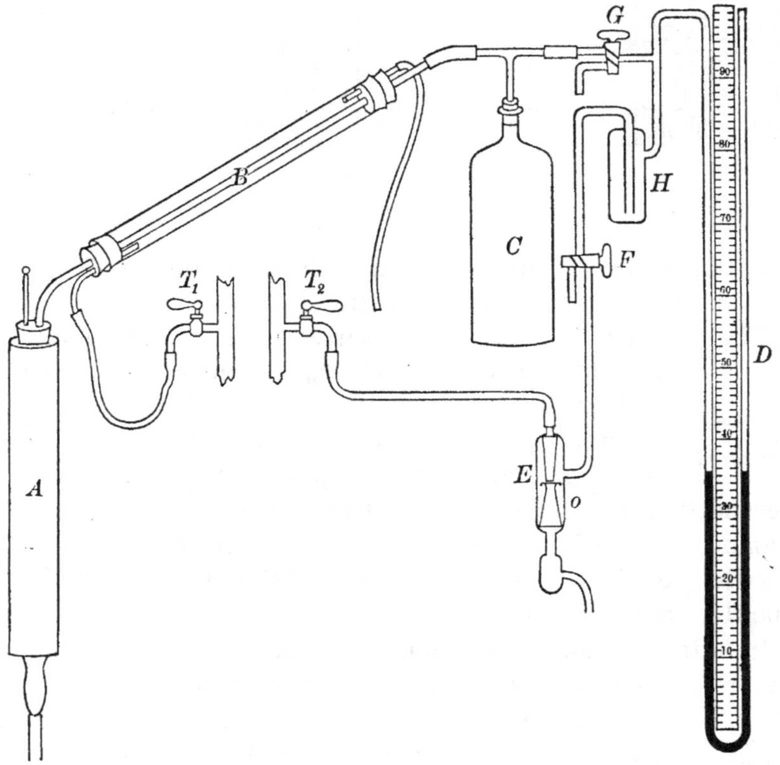

FIGURE 87

a slow current of water is passed from the tap T_1. It is only by virtue of the immediate condensation of the steam as it forms that the pressure within the boiler can be kept constant. C is an air-tight chamber of sufficient capacity (in this case about 4 liters) to prevent irregularities in the boiling from producing appreciable changes in the pressure. The only other essential features of the apparatus are a manometer D and any sort of an air pump. That shown in the figure is a Bunsen water pump, such as may be very conveniently used

The water pump.

in any laboratory which is supplied with running water. As soon as the tap T_2 is opened, a jet of water rushes through the orifice at *o* and draws with it the air from the chamber *E*, thus exhausting any vessel with which *E* is connected. *H* is a trap to prevent water from being sucked back into the manometer when the pump is stopped. *F* is a three-way stopcock (see Fig. 77b, p. 131) by means of which the boiler may be put into connection either with the pump or with the air, or cut off entirely from outside communication. *G* is also a three-way cock, which is inserted so that it may be unnecessary to disconnect the boiler if it is desired to use the water pump for exhausting purposes in other experiments.

Manipulation. First, start the circulation in the condenser, then turn *F* until *A* is in communication with the air, and start the water to boiling. After the conditions have become stationary, read the barometer and the boiling temperatures. Then turn *F* so that the boiler communicates with the pump, and allow the water to run until a difference of 5 or 10 cm. has been produced in the arms of the open mercury manometer *D*. In this regulation of pressure, strive to duplicate as nearly as possible some pressure used in the static method. Next close *F* entirely, stop the pump, and after waiting about two minutes take several observations of the new boiling point and the corresponding pressure. Then again start the pump, put the boiler into connection with it, and reduce the pressure to some second value used in **1**. Continue in this way until the boiling temperature has fallen to about 75°C. In this method the observations can not be conveniently carried to temperatures lower than 75°, because, with much further exhaustion, the difficulty of boiling with bumping is encountered. By attaching a bicycle pump to *F*, the boiling point for pressures somewhat higher than 76 cm. can be investigated. Correct the thermometer readings exactly as in **1**, and tabulate results in the form shown in the Record. Finally, it is required to plot in the note-book a full-page curve in which temperatures are represented by abscissae, and pressures by ordinates. The book values are to be indicated by dots, the values obtained in **1** by crosses, and those obtained in **2** by circles. The smooth curve, which comes as nearly as possible to touching all these points, is the curve required.

Record

1st Method						2d Method					Book values of pressures
Temperatures			Pressures			Temperatures		Pressures			
obs'd	corrected for (1)	corrected for (2)	obs'd	red'c'd to 0°	corrected	obs'd	corrected	barometer	manometer	total red'c'd	
						....	....	....	....	....	
						....	....	....	....	....	
						....	....	....	....	....	
						....	....	....	....	....	
						....	....	....	....	....	
						....	....	....	....	....	
						....	....	....	....	....	

Problems

1. Assume that the laws which hold for ideal gases hold also for vapors up to the very point of saturation, i.e., assume that equation (146) is applicable to saturated vapors. Then, with the aid of the known density of air at 0°, 76 cm., viz., .001293, the density of water vapor in terms of air, viz., .624, equation (146), and the above values for the pressures of saturated water vapor, calculate the densities of saturated water vapor at 10°C., at 40°C., at 70°C., at 100°C., and compare with the observed densities given in the table in the Appendix. The results will show how closely the gas laws apply even to saturated vapors.

2. In a uniform barometer tube in which the mercury stands but 40 cm. high, the space above the mercury is 40 cm. long, and contains at first only dry air. A few drops of ether are then introduced into the tube. If the tension of saturated ether vapor at the temperature of the room is 30 cm., find to what height above the mercury in the cistern the mercury in the tube will ultimately fall.

3. If the bulb of the apparatus shown in Fig. 85 were gradually heated above 100°, would any temperature ever be reached at which the water within the bulb would be observed to boil?

4. Explain why, from the standpoint of the kinetic theory, a lower temperature can be reached by fanning an open vessel of ether than by fanning an open vessel of water.

XIX

HYGROMETRY

Theory

Hygrometry is that branch of Physics which relates to the study of the water vapor contained in the earth's atmosphere.

The purpose of hygrometric observations. From the considerations present in Ex. XVIII, it is evident that, were it not for the presence of the air, the earth would always be covered with this vapor in a saturated condition, and precipitation in the form of fog, dew, or rain would accompany every fall in temperature, however slight. But the presence of air so retards the process of evaporation that even in the immediate neighborhood of lakes or oceans the condition of saturation does not usually exist. Hence it is, that precipitation often fails to occur even when the thermometer falls suddenly through many degrees. Nevertheless, a knowledge of the hygrometric state, i.e., the state of dryness, or wetness, of the atmosphere, or, what amounts to the same thing, a knowledge of the number of degrees through which the temperature must fall before dew can form, is of considerable importance not only for scientific but also for practical purposes, such, for example, as the forecasting of the probability of frost, or the maintenance within green-houses, drying-rooms, sick-rooms, and dwellings, of suitable climatic conditions.

The four hygrometric quantities sought. The four quantities involved in hygrometric determinations are:

1. The *density of the water vapor in the air*, i.e., the weight in grams of the water vapor contained in 1 cc. of space. This is usually called the *absolute humidity*. It is here represented by the letter d.

2. The *relative humidity* or the *degree of saturation*. This is represented by the letter r, and is defined as the ratio between the density of the water vapor existing in the atmosphere at any given time, and the largest density which it could possibly have at the

existing temperature; i.e., if D represent the density of saturated water vapor at the existing temperature, then the relative humidity r is given by $r = \frac{d}{D}$. Since the results of Problem **1**, page 162, have shown that at ordinary temperatures the density of saturated water vapor can be calculated with sufficient accuracy from the pressure which it exerts (obtained in Ex. XVIII), D may always be considered a known quantity (see also Appendix, table 6).

3. The *dew-point* τ, or the point to which the temperature must fall in order that the water vapor existing in the atmosphere may be in the saturated condition. Of course, as soon as the temperature falls below this point, condensation must ensue.

4. The *tension* or *pressure*, p, *which the water vapor in the air exerts at the existing temperature.*

As will presently appear, the experimental determination of any one of these four quantities taken in connection with the pressure-temperature curve of a saturated vapor (see Ex. XVIII), suffices for the calculation of all the rest.

The first attempt to construct an instrument for measuring hygrometric conditions was made about 1600, when an Italian named Sanctorius devised what is now known as an absorption hygrometer, an instrument usually associated with the name of de Saussure, a Genevan, who brought it into prominence in 1783. It is well known that many organic substances expand with an increase in the dampness of the atmosphere. De Saussure's hygrometer consisted of a human hair attached as in Fig. 88, so that changes in its length caused a pointer to move over a scale which was constructed by marking the position of the pointer in a saturated atmosphere 100, its position in a perfectly dry atmosphere 0, and then dividing the intervening space into 100 equal parts. The position of the pointer at any time was assumed to indicate directly the degree of saturation of the atmosphere (r). These instruments are still in common use, but they are now always graduated empirically by

Absorption hygrometers.

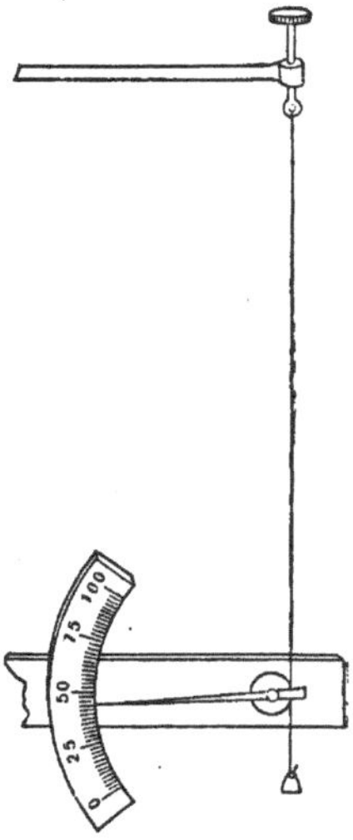

FIGURE 88

comparison with a dew-point hygrometer (see below); for careful experiments made by Regnault in 1845 showed that instruments constructed as above agree accurately neither with one another nor with the indications of dew-point hygrometers. Further, it is found that a given absorption hygrometer does not remain comparable even with itself for any long interval of time. Hence its indications can only be relied upon if it is frequently recalibrated.

Dew-point hygrometer.

Accurate measurements in hygrometry began with the introduction by the Englishman Daniell, in 1820, of the dew-point hygrometer, the essential principle of which had been employed by the Frenchman Le Roy as early as 1751. This instrument in one or another of its numerous modifications has become the standard of comparison for the testing and graduation of all other hygrometers. It consists essentially of a polished metal tube, the temperature of which is in some way lowered until dew is observed to form upon its surface. From this temperature of condensation τ, it is possible to determine all the other hygrometric constants.

To obtain the existing pressure of water vapor from the dew-point.

Thus the pressure p is obtained at once from τ and the pressure-temperature curve of a saturated vapor. It is simply the pressure of saturated water vapor at the temperature τ. For, although the cooling of the layers of atmosphere which are in contact with the metal surface causes an increase in the *density* both of the air and of the water vapor of which these layers are composed, yet, since the barometric pressure is in no way affected by the cooling, it is evident that the pressure both of the air and of the water vapor within these layers must remain precisely the same as outside, where no cooling takes place. The beginning of precipitation means only that in the layers adjoining the surface the density and pressure corresponding to saturation have been reached. If, then, the pressure within these layers is the same as outside, it is clear that table 6 gives the correct value of p.

Not so, however, with d. The table gives, indeed, the value of this quantity within the cooled layer, but the density at a distance is the density within the cooled layer multiplied by $\frac{T_\tau}{T_t}\left[=\frac{273^\circ + {}^\circ t}{273^\circ + {}^\circ\tau}\right]$; for when pressure remains constant, density is inversely proportional to absolute temperature. Instead of obtaining d in this way from the table, it may be directly calculated from p by the ordinary gas laws, viz., by (146). Thus, since the density of air at 0°, 760 mm., is .001293, and since under like condition of temperature and pressure the density of water vapor in terms of air is .624, (146) becomes, when d_{a_1}, p_1 and T_1 are replaced by d, p and T_t, aud d_{a_2}, p_2 and T_2 by .001293 × .624, 760 and 273,

To obtain the density of water vapor from the dew-point.

$$d = \frac{.001293 \times .624 \times 273}{760 \cdot T_t} p = \frac{.00029\, p}{T_t}, \qquad (149)$$

in which p must, of course, be expressed in mm. of mercury, since the barometric pressure has been so expressed. This extension to unsaturated vapor of the laws which hold rigorously only for ideal gases must be permissible in practice, since the results of Problem **1**, page 162, have shown that at ordinary temperatures equation (149) may be applied without appreciable error even to saturated water vapor. The relative humidity r can be at once obtained from either p or d and table 6; for

$$r = \frac{d}{D} = \frac{p}{P}.$$

One of the most perfect forms of the dew-point hygrometer is due to the Frenchman Alluard (1880), and is shown in Fig. 89. The nickel tube A, upon which the dew is formed, is about 2 cm. in diameter, and has one flat, highly polished side which is placed in close juxtaposition to a strip B of equally well polished nickel upon which no dew is formed. Tube A is filled with ether, the temperature of which may be lowered by causing a current of air to bubble through it. This is accomplished by means of an aspirator attached by a rubber tube to C. A bellows attached to F serves the purpose equally well. The experimenter

The Alluard form of dew-point hygrometer.

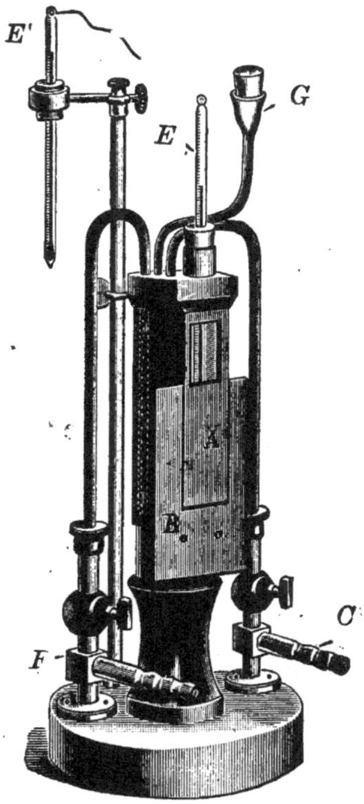

FIGURE 89

sits at a distance of 10 or 12 feet, so that the moisture from his breath or body may not affect the result, and observes the tube and thermometer through a telescope. At the instant at which *A* begins to look cloudier than *B*, he takes the temperature indicated by the thermometer *E*. He then stops the current of air and observes again the temperature at which the cloudiness disappears from *A*. With a little practice the temperatures of appearance and disappearance of the dew can be made to approach to within .1°C. The mean of these two temperatures is taken as the dew-point. This form of instrument should not be used in rapidly moving air, for then the layers of air which are in contact with *A* are removed before they can take up the temperature of the nickel, and in consequence the observed dew-points are too low.

The chemical hygrometer.

The indications of a dew-point hygrometer may be very nicely checked by means of the chemical method, first used for hygrometric determinations by the Swiss chemist Brunner in 1844. It consists in slowly drawing a known volume of air V through drying-tubes, preferably of anhydrous phosphorus pentoxide, and measuring the increase w in the weight of the tubes. Let v represent the volume of water which has been drawn out of the aspirator R during the experiment (see Fig. 90). The gas which has replaced this water consists of the perfectly dry air which has emerged from the drying-tubes and the water vapor which has formed from the water in R. Since this vapor may be assumed to be saturated, the pressure which it exerts is P, the pressure of saturated vapor corresponding to the temperature of the room. Hence the pressure exerted by the air alone which is within R is $H-P$, H being the barometric pressure. When this same air was outside, it exerted a pressure $H-p$, p being, as above, the pressure exerted by the water vapor in the outside air. Hence the volume which the air in R occupied before it entered the drying-tubes, i.e., the volume of air V in which the weight

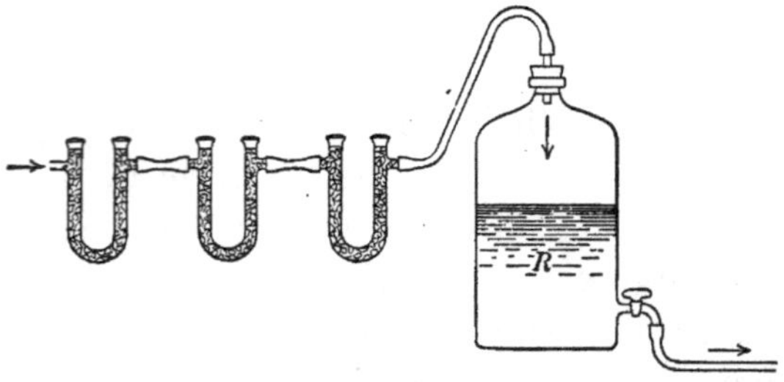

FIGURE 90

w of water vapor was contained, is given by (see Boyle's Law)

$$\frac{V}{v} = \frac{H-P}{H-p}. \qquad (150)$$

From (149), (150), and the relation $\frac{w}{V} = d$, there results

$$w = \frac{.00029\,p}{T_t} \cdot \frac{H-P}{H-p} \cdot v, \qquad (151)$$

an equation which contains only one unknown quantity, viz., p. After p has been determined from (151), d and r may be found as above, while τ is taken from table 6; i.e., it is the temperature of saturation of water vapor corresponding to the pressure p. In case P and p have nearly the same value, (150) gives $V = v$, in which case d is obtained at once from $d = \frac{w}{v}$, and p is then found from (149).

This chemical method leaves nothing to desire in the matter of accuracy, but since an observation usually requires from 1 to 3 hours, the result is, of course, only a mean value of the humidity during this time. It is little employed in practice.

The wet-and-dry bulb hygrometer.

The instrument which is now most extensively used in meteorological observations was first conceived by the Scotch physicist John Leslie in 1790. It was given its present form by August, in Berlin, about 1825. It is called the wet-and-dry bulb hygrometer, and consists of two sensitive thermometers mounted side by side, one of which has its bulb wrapped in muslin, which is kept moist by means of a cotton wick immersed in a water reservoir c (see Fig. 91). The temperature t' indicated by the wet-bulb thermometer is always lower than the temperature t shown by its dry-bulb neighbor, unless the water vapor in the air is already in a state of saturation, for the evaporation which otherwise takes place produces cooling (see page 155). The amount of this cooling evidently increases with the rapidity of evaporation, which is in turn directly proportional to the dryness of the atmosphere,

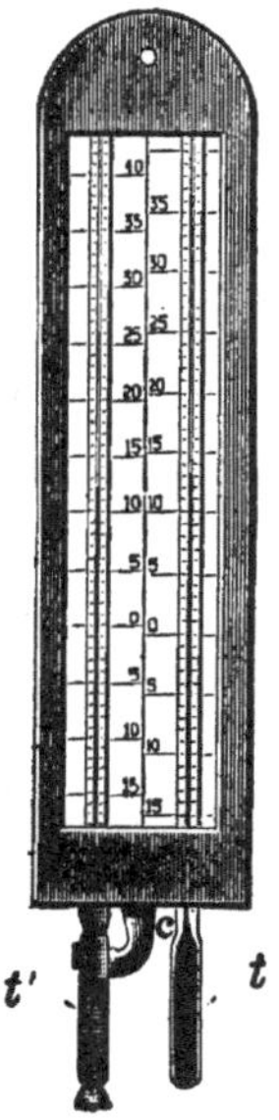

FIGURE 91

i.e., to $D-d$, or, what amounts to the same thing, to $P-p$. But it also depends upon the density of the air into which the evaporation takes place, i.e., upon the barometric height H, as well as upon the velocity of the air currents (see page 155). A long series of comparisons of this instrument with the dew-point hygrometer has given for air in moderate motion the empirical formulae

(for t' above 0°) (for t' below 0°)

$$p = P - 0.00080\,H\,(t-t'), \quad \text{and} \quad p = P - 0.00069\,H\,(t-t'), \qquad (152)$$

in which p, P and H are expressed in mm. of mercury and $(t-t')$ in degrees centigrade. When p has been determined from this equation, d is given by (149), and then r and τ from table 6. For air at rest, the numerical quantities in (152) are considerably too small. In order to make (152) applicable to indoor observations, it is recommended to hang the instrument from a long cord and to let it swing as a pendulum for several minutes before taking the readings. Hygrometer makers usually specify carefully the conditions under which instruments of this type are to be used, and furnish with them empirical tables.

Experiment

Object. To compare the indications of a dew-point and a wet-and-dry bulb hygrometer.

Directions. Place the Alluard hygrometer in a room which is free from evaporating water, pour ether into G (Fig. 89) until the liquid surface is above the window in A, turn the polished metal face into as favorable a light as possible, set the observing-stand, telescope, and aspirator at a convenient distance, and take, first, a rough observation of the dew-point. In subsequent observations regulate the evaporation of the ether so that the temperature falls very slowly in the neighborhood of the point sought, and take the reading of the thermometer E when the first cloudiness begins to show upon A. In taking observations with a rising temperature, an occasional bubble may be allowed to pass through the ether in order to keep it stirred. Tabulate results as in the Record. Then bring into the room a wet-and-dry bulb hygrometer, and take a set of observations in the manner indicated in the Theory.

Record

	Dew-Point Hygrometer				Wet-and-dry bulb
No. of obs'n	Room tem. (E′)	Dew app'd (E)	Dew disapp'd (E)	Mean	$H =$ ——
1	——	——	——	——	$t =$ ——
2	——	——	——	——	$t' =$ ——
3	——	——	——	——	$\therefore p =$ ——
4	——	——	——	——	$\therefore d =$ ——
5	——	——	——	——	$\therefore r =$ ——
	$\therefore p =$ ——	$\therefore d =$ ——	$\therefore r =$ ——	$\tau =$ ——	$\therefore \tau =$ ——

Problems

1. When the relative humidity is .47 at 21°C., what will be the dew-point?

2. If the temperature of the air at sundown on a clear day be 10°, and if the wet-bulb thermometer read 8°C., at what temperature will dew form? Need there be fear of frost during the night? (Bar. ht. 750 mm.)

3. If the wet-bulb thermometer of Problem **2** had read 4.5°C., what would have been the dew-point? In this case frost would have been almost certain. Why?

4. Dry air at 18°, 755 mm., weighs .001205 gm. per cc. Find the density of the atmosphere at this temperature and pressure when the dew-point is 10°C.

XX

ARCHIMEDES' PRINCIPLE

Theory

The law which asserts that *the loss in weight experienced by any body when immersed in a fluid is equal to the weight of the displaced fluid*, was discovered by the immortal Greek philosopher Archimedes, who perished in the siege of Syracuse in 212 B.C. Unlike many of the laws which have preceded, it is not an approximation, nor is it primarily empirical, experiment having only served to confirm results which follow with certainty from theory. The work of Archimedes was not known in the middle ages, and the law was rediscovered in 1586 by the Flemish scientist Stevin, who advanced for it the following proof: Within a body of fluid, isolate in thought some mass by means of any imaginary bounding surface S (see Fig. 92). Since the mass of liquid within this boundary is in equilibrium, its weight must be neutralized by forces whose existence is due to the surrounding liquid. But these forces which are exerted by the surrounding liquid upon the surface S depend only upon the conditions which exist outside of S, and are wholly independent of the nature of the substance within S. Hence any immersed body whatever which has the surface S must be buoyed up by forces the resultant of which is equal to the weight of the displaced liquid.

Proof of law of Archimedes.

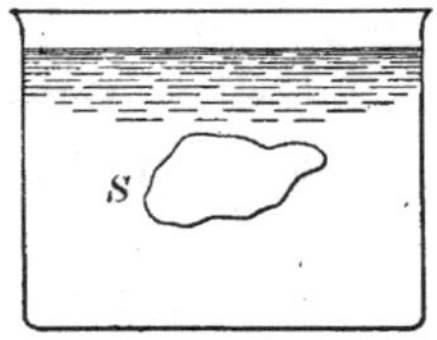

FIGURE 92

It follows from this law that when a body is balanced upon scales *in air*, the balancing weights do not accurately represent the true weight of the body, i.e., its weight in vacuo. For both the body and the weights are buoyed up by the air, and since in general the volume of air displaced by the body is not the same as that displaced by the weights, the buoyant effects upon the two sides must be different. However, with the aid of the "principle of moments" the true weight X can be easily obtained from the apparent weight W (i.e., the weight of the weights), the volume

Application of the law of Archimedes to the determination of a correct weight.

V of the body, the volume V' of the weights and the density σ of air. For, since the resultant downward force on the side of the body is $(X - V\sigma)$ grams, and that on the other side $(W - V'\sigma)$ grams, the equation of balance for equal balance arms becomes

$$X - V\sigma = W - V'\sigma \quad \text{or} \quad X = W + (V - V')\sigma. \qquad (153)$$

Hence the so-called *air correction*, i.e., the correction which must be applied to the apparent weight W in order to obtain the true weight X, is simply the difference between the weight of air displaced by the body and that displaced by the weights. This correction is evidently positive if $V > V'$, negative if $V < V'$. Since the density of the weights is always known (for brass it is 8.4), it is usually convenient to replace V' by $\frac{W}{8.4}$, and to write (153) in the form

$$X = W + \left(V - \frac{W}{8.4}\right)\sigma. \qquad (154)$$

Or again, if the density d of the body happens to be roughly known, (153) takes the approximately correct form (since, on account of the smallness of the correction term, X is usually very nearly equal to W),

$$X = W + \left(\frac{W}{d} - \frac{W}{8.4}\right)\sigma. \qquad (155)$$

In deducing the above expressions for a correct weight X, the balance arms were assumed to be equal. Although this is, of course, seldom rigorously the case, it was shown on p. 39 that the error arising from any inequality can be completely eliminated by taking a mean of the weighings made on both pans. Hence, in order to obtain a very accurate weight, it is necessary first to find W by means of a double weighing, and then to apply to W the air correction as shown in (153).

In case the quantity sought is a small increase or decrease in weight, as in Exs. XV and XVII, the rigorous process described in this section is not to be recommended; for neither the error in the balance arms nor the air correction due to the weights is in such cases likely to be an appreciable quantity. For most purposes single uncorrected weighings, in which, however, the body weighed always hangs from the same balance arm (see p. 118), are sufficiently accurate.

Archimedes' principle also furnishes the most convenient and accurate method of determining densities. For, if any body, regular or irregular, whose absolute weight is X grams, be found to lose L grams when weighed in water of density ρ, it is evident at once from the statement of Archimedes' principle that the volume of the body is $\frac{L}{\rho}$, and hence that its density d, which is by definition $\frac{\text{mass}}{\text{volume}}$, is given by

Application of the law of Archimedes to the determination of the density of a solid.

$$d = \frac{X}{\frac{L}{\rho}} = \frac{X\rho}{L}. \qquad (156)$$

But, in order to find L accurately, an air correction of the ordinary form must be applied to the apparent loss of weight. For let the body whose true weight is X and whose apparent weight is W be immersed in water and weighed, first when hung from one balance arm, then from the other, and let this mean apparent weight in water be W_1. The equation of balance for this case, in which the body hangs in water while the weights hang in air, is evidently

$$X - V\rho = W_1 - \frac{W_1}{8.4}\sigma. \qquad (157)$$

Substituting the value of X found in (154), there results at once

$$V\rho = L = (W - W_1) + \left(V - \frac{W - W_1}{8.4}\right)\sigma, \qquad (158)$$

which is an equation of the same form as (154), the apparent weight W having been simply replaced by the apparent *loss* of weight $(W - W_1)$.

The very statement of Archimedes' principle also suggests its use for determining the densities of liquids. For, if L represent the loss of weight of a body in water of density ρ, and L' the loss of weight of the same body in another liquid of unknown density d, then, since the volume of the body is $\frac{L}{\rho}$, the density of the unknown liquid is given by

Application of the law of Archimedes to the determination of the density of a liquid.

$$d = \frac{L'}{\frac{L}{\rho}} = \frac{L'\rho}{L}. \qquad (159)$$

The weighings in the two liquids may, of course, be made with an ordinary balance, precisely as above, but for the sake of rapidity a modified form of balance due to Mohr (see Fig. 93) is commonly used.

The principal features of this balance are (1) the division of one arm into ten equal parts, and (2) the use of weights of convenient shape to hang from any of the ten notches, the tenth of which is the hook *c*. Since each weight can be given ten different values, a very small number of weights is required. There are never more than five. Further, as will presently appear, the absolute value of these weights is of no importance if only their ratios are accurately represented by the numbers 1, .1, .01, and .001 (the fifth weight is a duplicate of 1, used for the sake of extending the range of the instrument). For, suppose that the body *B* is of such weight that, when hung in air from the hook *c*, it is possible by means of a little adjustment (see Experiment) to balance the beam so that the two points at *a* are exactly together. Next suppose that when *B* is immersed in water as in the figure, it is necessary, in order to bring the points together again, i.e., in order just to counterbalance the buoyancy of the water, to hang weight 1 and weight .1 from the hook *c*, and weights .01 and .001 from notch 4. The loss of weight *L* of the bulb, expressed, not in grams, but in terms of weight 1, is then evidently 1.1044. Suppose now that when the water is replaced by another liquid of the same temperature, it is found necessary, in order to bring the points again together, to hang 1 from notch 9, .1 from notch 2, .01 from notch 4, and .001 from notch 9. The

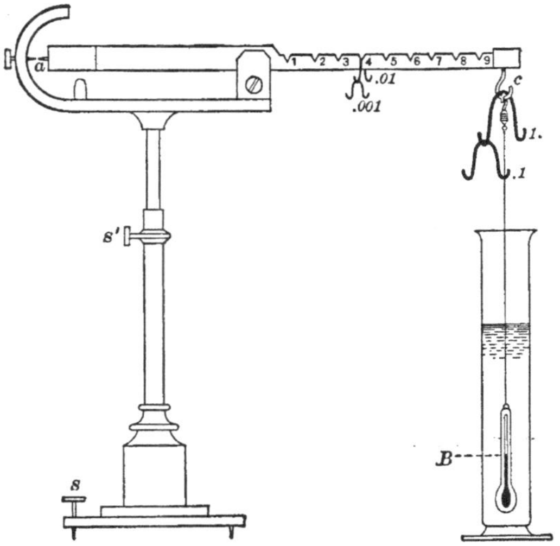

FIGURE 93

loss of weight of the bulb in this liquid, expressed as above in terms of 1, is then .9249. Hence (159) gives

$$d = \frac{.9249}{1.1044}\rho.$$

If ρ is known d is at once obtained, no matter what happens to be the weight of 1. In practice this weight is usually arranged to be as nearly as possible equal to the weight of the water displaced by the bulb at 15°C. In this case the reading for water at this temperature is evidently 1.0000, and if no high degree of accuracy is required, the reading of the instrument when the bulb is immersed in the unknown liquid gives at once the density of that liquid. Rigorously the apparent losses in weight L and L', obtained by means of a Mohr's balance, are subject to air corrections, but since in practice the balance is only used to compare densities which differ comparatively little from one another, the influence of these corrections upon the result is negligible. They could be made, if necessary, by reducing L and L' to grams and then applying (158), V being in this case the volume of the bulb.

Experiment

Object. **1.** To compare a density determination made upon an accurately turned cylinder of aluminium by means of weight and dimension measurements, with a determination made upon the same cylinder by means of observations of weight and loss of weight in water. **2.** To compare the results of measurements upon the density of a liquid made by a Mohr's balance, with those made by an ordinary constant-weight hydrometer.

Volume. DIRECTIONS.—**1.** First find the volume of the cylinder from very careful measurements of its dimensions, made by means of calipers.

Then suspend the cylinder from a very fine platinum wire and make an absolute weighing in air of the cylinder and wire as follows:

Absolute weighing. (1) Choose any convenient zero to which to refer your weighings, e.g., the 10 mark.*

*The true zero of the unloaded balance may be determined if desired, but this is never necessary with a double weighing; for evidently, if the arbitrarily chosen zero differs from the true zero, the weighing upon one

(2) Hang the body from the left arm and, using all the precautions mentioned on page 116, find the resting point R_1, which corresponds to such a weight in the right pan as will keep the pointer swinging within three or four divisions of the chosen zero.

(3) Add a 2 mg. weight to the lighter side and take the new resting point R_2; then at once raise the arrest and compute the sensitiveness.

(4) Exchange the positions of the cylinder and the weights, at the same time making and recording a careful count of the latter, exclusive of the two added milligrams.

(5) Find the new resting point R_3; then raise the arrest and count again the weights as they are returned to their respective compartments in the box of weights.

(6) Calculate from the sensitiveness the corrections necessary to apply to the counted weight upon each side, in order to make both of the resting points coincide with the chosen zero. Thus, if the zero is 10, the sensitiveness 2.1, and the resting point when the cylinder is on the right pan 10.9, the correction, in this case to be subtracted from the counted weight, is $\frac{10.9-10}{2.1} = .4$. Represent the mean corrected weight by W and the result obtained by applying to W the air correction (154) by X.

Next immerse the aluminium cylinder, in the manner shown in Fig. 94, in a beaker of distilled water which has been recently freed from air by boiling, but which has again regained the temperature of the room. Carefully remove from the immersed body all bubbles, even the smallest, by means of a camel's hair brush, then weigh on both sides, following exactly the directions 1 to 6 above. Represent the mean weight in water by W_1, and let the result obtained by applying the air correction (156) to the apparent loss in weight, $W - W_1$, be represented by L.

Weighing in water.

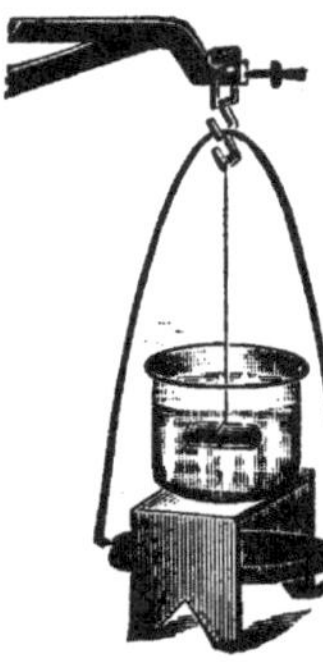

FIGURE 94

side will be just as much too large as that upon the other is too small, and the mean will therefore be the same. If, however, a weighing is made upon balances which are known to have arms so nearly alike that a double weighing is unnecessary, this single weighing must be referred to the true zero obtained by the method of oscillations (see p. 116).

In order to obtain $\bar{X}$, the weight of the cylinder alone, the weight w of the suspending wire must, of course, be obtained and subtracted from the joint weight X obtained above. Since this is a very small quantity, a single weighing made upon one pan, and uncorrected for displaced air, is sufficient. To make this weighing, find:

Weighing the suspending wire.

(a) The true zero of the balance (see note, p. 176);

(b) The resting point r, when the wire alone is on one pan, and some nearly equal weight on the other;

(c) The resting point r_2 (after 2 mg. have been added to determine the sensitiveness for this load);

(d) The weight w of the wire (obtained from the weight in the pan and the sensitiveness).

The loss of weight L represents, of course, the weight of water displaced both by the immersed cylinder and the immersed portion of the suspending wire. Estimate roughly the volume v of this immersed wire by measuring its diameter δ (with the micrometer caliper) and its approximate length l. This volume v is approximately the weight of the displaced water. Hence the loss of weight $\bar{L}$ of the cylinder is given by $\bar{L} = L - v$.

Correcting L for immersed wire.

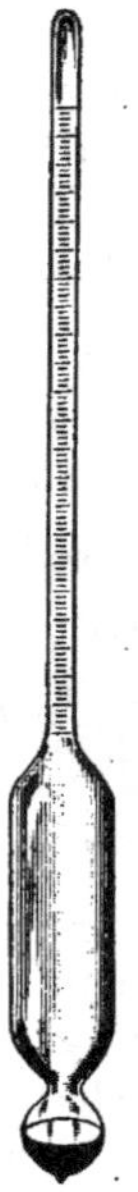

FIG. 95

2. Loosen set screw s' of the Mohr's balance (Fig. 93), and turn the base until the leveling screw s lies in the vertical plane which includes the beam. Adjust vertically at s' until a convenient height for immersion is reached. Then from the hook c hang the float B, in air, and level by means of s until the two points at a are very accurately together. Then, by means of the weights bring the points again together, first when the float is immersed in distilled water, then when immersed in the liquid of unknown density, and take the corresponding readings (see Theory). The density of the water at the observed temperature is taken from the table of water densities. The temperatures of the two liquids compared must be the same, otherwise a correction must be applied because of the expansion of the float.

Compare the density given by the Mohr's balance with the indication of a direct-reading, constant-weight hydrometer (see Fig. 95). The theory of this instrument is too simple to require explanation. The reading is made through the liquid, the eye being placed as little as possible beneath the level of the surface. If the instrument does not read the correct density of distilled water at the observed temperature, a correction amounting to the difference must be applied to its indication of the density of the other liquid. For very accurate density determinations with hydrometers of this sort it is important that the stem be wet above the point of contact with the liquid, since otherwise the capillary forces between the liquid and the stem may give rise to very considerable errors. Hence, before taking a reading, push the instrument down below its natural position of equilibrium and then let it rise. If in this operation the liquid be observed to be depressed about the stem, instead of elevated, the stem should be carefully cleaned with an alkali, e.g., potassium hydrate.

Record

1. Bar. ht. = —— Temp. of room = —— ∴ Density of air = ——

Diam. of cylinder 1st obs. —— 2d —— 3d —— 4th —— mean = ——

Height of cylinder " —— " —— " —— mean = —— ∴ V = ——

Weighing of	cyl. + wire in air	cyl. + wire in water		wire alone		
Chosen zero	= ——	——	Zero	= ——	t *	= ——
R_1 (cyl. left)	= ——	——	Rest. pt. r_1	= ——	ρ	= ——
R_2 (cyl. left)	= ——	——	Rest. pt. r_2	= ——	l	= ——
R_3 (cyl. right)	= ——	——	∴ Sensitiv's	= ——	δ	= ——
Counted wts.	= ——	——	Counted wt.	= ——	v	= ——
∴ Sensitiv's (S)	= ——	——	Cor'd wt. (w)	= ——		
Cor'd wt. left	= ——	——	∴ $\overline{X} (= X - w)$	= ——		
Cor'd wt. right	= ——	——	$\overline{L} (= L - v)$	= ——		

Mean (W) = —— W_1 = —— $d = \frac{\overline{X}}{V}$ = —— $d = \frac{\overline{X}\rho}{\overline{L}}$ = ——

True wt. (X) = —— L = —— % difference in d's = ——

2. Reading of Mohr's balance in water = —— tem. = —— ρ = ——

Read'g of Mohr's balance in unknown liq. = —— tem. = —— ∴ d = ——

Hydrom'r in water = — ∴ corr'n = — in liquid = —— ∴ d (cor'd) = ——

---

* t = tem. and ρ = den. of water. l, δ, and v are length, diameter, and volume of immersed portion of wire.

Problems

1. If the length measurements made upon the cylinder were 3.021, 3.023, and 3.024 cm., and if the diameter measurements were 2.567, 2.563, 2.564, and 2.562 cm., find what is the first uncertain figure in the number which represents the volume. (Results should never be recorded farther than to one place beyond the first uncertain figure.)

2. If the weighings can all be made with such accuracy that the tenths mg. place is the first place of uncertainty, find to how many more places of certainty the density is given by the loss of weight method than by the direct measurement method (weight of cylinder about 12 gm.).

3. Archimedes discovered his principle when seeking to detect a suspected fraud in the construction of a crown made for the tyrant of Syracuse. It was thought to have been made from an alloy of gold and silver instead of from pure gold. If the crown weighed 1000 gm. in air and 940 gm. in water, find how many gm. of gold and how many of silver were used in its construction.

Assume that the volume of an alloy is the combined volumes of the components, and take the density of gold as 19.3 and that of silver as 10.5.

4. A 10-gm. weight placed upon a block of wood weighing 30 gm. sinks it to a certain point in water. In a salt solution it requires 15 gm. to sink the wood to the same point. Find the density of the salt solution.

5. If the density of ice is .918 and that of sea water is 1.03, find what fraction of the total volume of an iceberg is above water.

6. A cylinder of cork 10 cm. high and of density .2 floats upon water. If the air above the water be removed, will the cork sink or rise in the liquid? How much?

Assume incompressibility in both cork and water.

7. Suppose that a constant-weight hydrometer which it is desired to calibrate is immersed in two liquids whose densities are known to be 1. and 1.1, that the two points of immersion are accurately marked, and that the intervening stem is then divided into 10 equal parts. Assuming that the stem is accurately cylindrical, will this hydrometer give correct readings in liquids of intermediate densities? Why?

XXI

CAPILLARITY

Theory

One of the fundamental assumptions made in elementary hydrostatics is that a liquid, like so much sand, exerts pressure merely by virtue of its weight, and by virtue of the property, which it possesses in common with all fluids, of transmitting pressure in all directions.* Thus if p_0 be the pressure (force in grams per unit area) exerted upon the surface of a liquid of density d, then the number of grams of

Ordinary law of liquid pressure.

*This property follows at once from the fact of fluidity and the fundamental laws of mechanics. Thus let A (Fig. 96) be a substance concerning which the one assumption is made that it is capable of adjusting itself with perfect ease to any change in the shape of the containing vessel. Let f and f' be two forces acting upon frictionless pistons 1 and 2 of areas a and a' respectively. It is required to find the ratio which must exist between the forces f and f', in the condition of equilibrium. Let piston 1 move uniformly down a distance s thus crowding out of cylinder 1 a volume of fluid as. f' must then move uniformly up a distance s' such that $as = a's'$. But from the "principle of work" (scholium to Third Law) in the condition of equilibrium (rest or uniform motion) $fs = f's'$. Hence $\frac{f}{f'} = \frac{a}{a'}$, i. e., the forces which must act on the pistons in the condition of equilibrium are directly proportional to their areas. Since the directions of the forces f and f' are wholly arbitrary, there results the law first announced in 1653 by the French philosopher, mathematician, and man of letters, Pascal,—"*The forces transmitted by fluids act equally in all directions and are proportional to the areas of the surfaces upon which they act,*"—a law which finds its most beautiful experimental demonstration in the hyrdaulic press.

Pascal's Law.

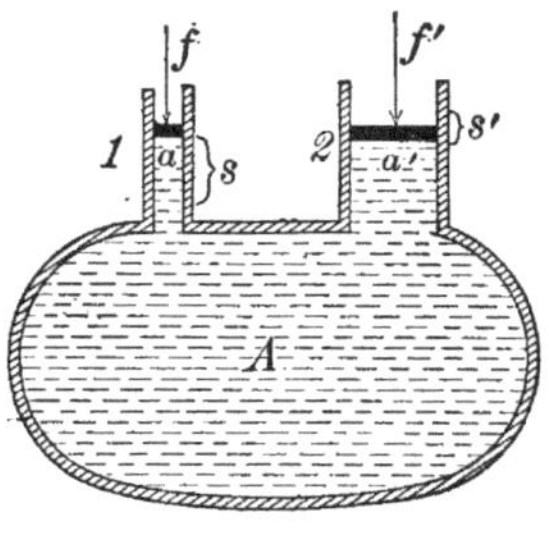

FIGURE 96

pressure P, which exists at any depth z beneath the surface, is given by (see Fig. 97)

$$P = p_0 + zd. \qquad (160)$$

There follows at once then the result, in general confirmed by experiment, that a liquid contained in a series of communicating vessels must take the same level in all of them, no matter how different they may be in size or shape.

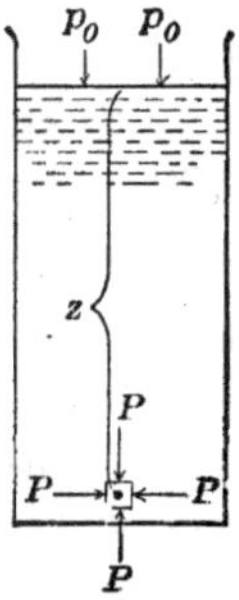

FIGURE 97

But it was observed as early as 1500 by that most universal of all geniuses, Leonardo da Vinci, that when a tube approaches capillary dimensions water rises in it far above the level in the outside vessel (see Fig. 101). Later and more careful investigation has shown the existence of a large number of different phenomena to which the ordinary laws of hydrostatics do not apply. These are usually all called "capillary phenomena," because they were first observed in connection with capillary tubes. They are all manifestations of those intermolecular forces which were assumed in Ex. XVIII in order to reconcile the existence of liquid surfaces with the theory of molecular motion. This section is devoted to a study of the effects of these forces; and since these effects would of necessity be just the same whether the molecules are at rest or in motion, the fact of motion will for the present be disregarded.

Capillary phenomena.

The simplest experiments place the existence of these intermolecular forces beyond the possibility of doubt, and show at the same time that, while they have enormous values at short range, they diminish so rapidly with the distance as to become wholly inappreciable at distances which still amount to extremely minute fractions of a millimeter. Thus a drop of mercury, instead of spreading out into an infinitely thin layer, as it would do if gravity alone acted upon its molecules, is held together in globular form. A sheet of glass may be brought extremely close to a surface of water without appearing to be attracted toward it in the slightest degree, but as soon as contact is made the glass clings to the water with remarkable tenacity. The surface of two metal blocks may be brought to within a thousandth of an inch without showing any appre-

Evidence as to the existence and nature of molecular forces.

ciable attraction, but as soon as they are brought somewhat nearer, as by pressure or welding, it requires tons of weight to pull them apart again. The operation of the aspirator pump described in Ex. XVIII is due to the attraction between the air about the orifice *o* and the outpouring current of water.

Laplace's theory of molecular pressure.

Starting, then, with these two facts, (1) the existence of intermolecular forces, and (2) the rapid diminution of these forces with the distance,* the great French geometrician, Laplace, first developed, about 1807, a theory of capillary action. His reasoning was somewhat as follows: Let *r* represent the distance within which one molecule attracts another with a force which is large enough to deserve consideration. Laplace called it the *radius of influence of molecular force.* It is, of course, not a quantity the magnitude of which is definitely fixed, but it probably never exceeds .00005 mm.

Now a molecule *m'* in the interior of a liquid is indeed acted upon by all the multitude of molecules lying within a sphere of radius *r* (see Fig. 98); but, by virtue of symmetry, the resultant of all these forces is evidently zero; so that *m'* may be treated as though it were under the influence of no molecular force whatever. Not so, however, with any molecule *m* which is nearer to the surface than the distance *r*. For while the molecules within the space *cefd* exactly neutralize the effects of the molecules within the space *acdb*, the downward resultant of the forces of all the

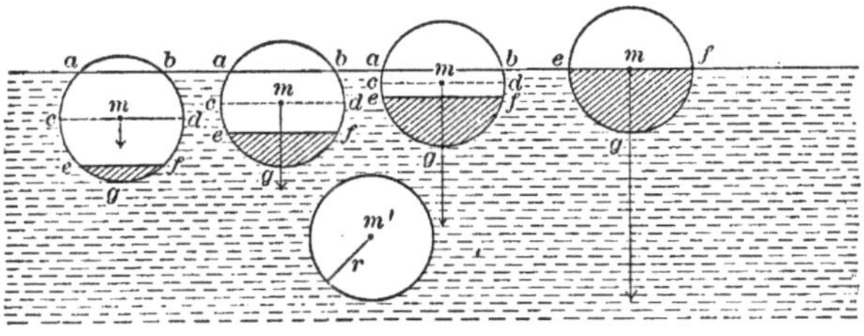

FIGURE 98

*In order to account for this rapid diminution with the distance it is not necessary to assume that these intermolecular forces are any other than those concerned in ordinary gravitation. For the law of inverse squares can be considered to hold for the attractions of masses of finite volume only so long as the distance between the nearest points of the attracting bodies is infinitely large in comparison with the distances between the molecules of the bodies.

molecules in *efg* is wholly unbalanced. This force continually urges *m* into the interior of the liquid. But all the other molecules in the same horizontal layer with *m* are urged inward with the same force and all the molecules in other layers which are within a distance *r* of the surface are urged in with other forces. The result of all these unbalanced forces acting upon all the molecules contained in the surface layer of thickness *r* (called the *active layer*) must be, then, an interior pressure of uncertain, perhaps enormous, magnitude. It has been estimated for water at something like 10,000 atmospheres, but it has never been measured directly and never can be. For since a liquid is always bounded on all sides by a surface, this molecular pressure usually balances itself and therefore cancels out in hydrostatic measurements. Hence it is that equation (160), which leaves the existence of molecular pressure altogether out of account, and treats the liquid molecules as though they were so many independent grains of sand, nevertheless gives, in general, correct results. But it is exactly such apparent violations of the ordinary hydrostatic laws as are shown in capillary phenomena, which furnish a beautiful proof of the existence of Laplace's molecular pressure.

Variation of molecular pressure with curvature of surface.

For it is easy to show that this pressure must be greater underneath a convex, and less underneath a concave surface, than it is beneath a flat one. Thus, let *M* (Fig. 99) represent a molecule which lies in the active layer at a given distance beneath a surface, and let the circle drawn about *M* represent the sphere of influence of molecular forces. The surface will first be assumed to be flat (*acb*), then convex (*ecf*), and then concave (*gch*). In the first case, since *pidjq* neutralizes *pacbq*, the resultant downward force acting upon *M* is due to the attraction of the molecules lying within the segment *iojd* of the sphere. In the second case *pkdlq* neutralizes *pecfq*, and the resultant downward force is due to *kold*, a volume which is greater than *iojd*. In the third case the resultant force is due to *mond*, a volume which is less than *iojd*. Hence the resultant downward force upon the molecules in the active layer is greatest beneath the convex, and least

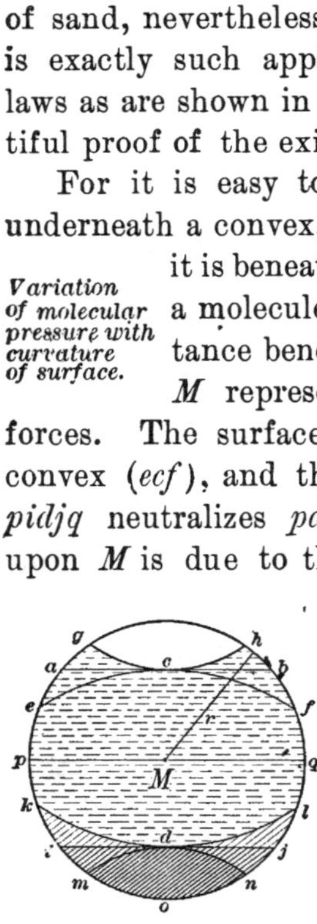

FIGURE 99

beneath the concave, surface. It is evident also from the same kind of reasoning that the greater the convexity, the greater this pressure. Thus, if the pressure beneath a plane surface be represented by P_0 (Laplace named this the *normal pressure*), that beneath a curved surface is $P_0 \pm p$, in which the magnitude of p depends upon the nature of the liquid and the magnitude of the curvature, while its sign is + or − according as the surface is convex or concave. Laplace proved by a mathematical analysis of the forces exerted by segments of the kind shown in Fig. 99, that p, expressed in terms of a characteristic constant A of the liquid (called its capillary constant), and the two principal radii of curvature R and R' of the surface, is

$$p = A\left(\frac{1}{R} + \frac{1}{R'}\right). \tag{161}$$

This makes the ascension or depression of liquids in capillary tubes perfectly intelligible. For, starting with the fact of observation (accounted for below) that a liquid in a small tube assumes a curved instead of a flat surface, its rise or fall in the tube, according as the surface is concave or convex, follows as a matter of course from a very simple consideration of the pressures involved.

Explanation of capillary ascension or depression.

Thus, the correct value of the pressure at a distance z below a plane surface is not $p_0 + zd$, as assumed in (160), but rather $p_0 + P_0 + zd$, and the pressure at the same distance z beneath a concave surface in a capillary tube (see Fig. 100) is

$$p_0 + \left[P_0 - A\left(\frac{1}{R} + \frac{1}{R'}\right)\right] + zd.$$

Hence from Pascal's Law of the equal transmission of pressure (see note, p. 181) there can be no equilibrium until the stronger molecular pressure beneath the flat surface has pushed up the liquid in the tube to such a height h (see Fig. 101) that the total pressures

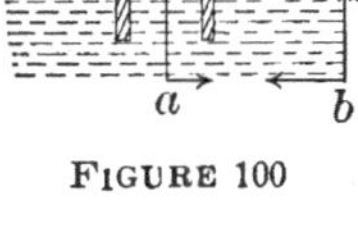

FIGURE 100

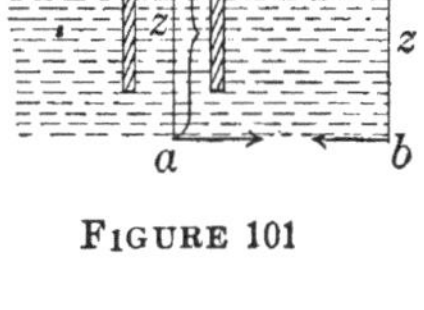

FIGURE 101

at any two points a and b in the same horizontal plane are the same, i.e., until

$$p_0 + P_0 + zd = p_0 + \left[P_0 - A\left(\frac{1}{R} + \frac{1}{R'}\right)\right] + zd + hd;$$

or

$$hd = A\left(\frac{1}{R} + \frac{1}{R'}\right). \qquad (162)$$

R and R' are the curvatures at the point considered, e.g., 1, 2, or 3 (Fig. 101), and h is the elevation of this point above the outside plane surface.

It thus appears that, correctly speaking, a liquid does not rise in a capillary tube because of a capillary *attraction*, any more than it rises in a suction pump because of the attraction of the vacuum created by the lifting of the piston. In both cases the liquid is *pushed* up by a pressure existing outside. In the case of the pump this is the atmospheric pressure acting on top of the water in the cistern; in the case of the capillary tube it is the normal pressure P_0 acting in the surface layer of the outside liquid.

If it were possible to remove entirely the molecular pressure within the tube, the height of rise would be a measure of P_0, just as the height of rise of the water in a long tube from which the air is entirely removed, is a measure of the atmospheric pressure. Since, however, nothing more can be done than to obtain a curved surface within the capillary tube, it is only the capillary constant A which can be found from observations of the height of ascension h, the density d, and the radii of curvature R and R' [see (162)].

Measurement of the capillary constant.

In the general case it would be difficult to measure R and R', but if the tube is cylindrical, then it follows from symmetry that at the middle of the meniscus $R = R'$, and (162) reduces to

$$hd = \frac{2A}{R}. \qquad (163)$$

If, farther, the tube is so small that the height of the meniscus m (see Fig. 101) is negligible in comparison with h, i.e., if h is practically constant for all points of the surface, then it follows from (162) that the curvature $\left(\frac{1}{R} + \frac{1}{R'}\right)$ is also practically constant. But the only surface of constant curvature which can fulfill the condition imposed by (163) is a section of a sphere. If finally, then, the liquid can be made to *wet* completely the interior of the tube, so that its angle of contact α with the walls is 180°, then the meniscus must be a hemisphere, and the radius R is simply the radius of the tube.

Equation (163) is then applicable to all cases for which these conditions hold.* It shows that the height of rise h is inversely proportional to the radius of the capillary—a law discovered experimentally by an Englishman in 1718 and called after him the law of Jurin. Equation (163) thus makes the measurement of the capillary constant a very simple matter in the case of liquids which wet solids of which capillary tubes can be made.

Another interesting conclusion which can be drawn from the above reasoning is that, since in equilibrium the height h depends only upon the curvature and the density, the dimensions of the capillary above or below the point of contact have no effect whatever upon the phenomena. Thus, if water be drawn up into tubes of such different shape as a and b (Fig. 102), it should come to rest in the descent at precisely the same level in both. This conclusion is wholly confirmed by experiment.

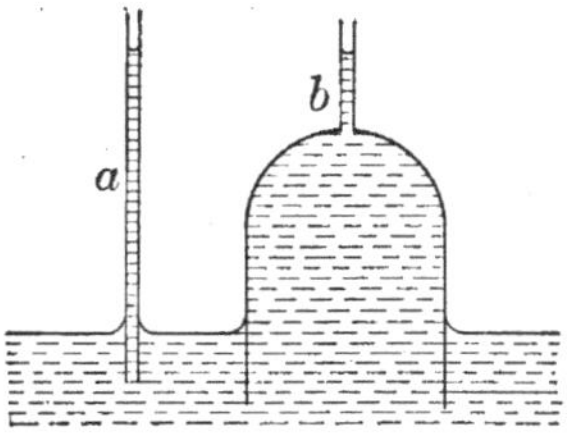

FIGURE 102

It only remains to show why a liquid in a capillary tube assumes a curved surface—a task of no difficulty when it is remembered that a liquid surface can be in equilibrium only when it is perpendicular to the resultant force acting upon its molecules. This property follows simply from the fact of mobility of the particles. For, if the force acting upon the surface molecules had any component parallel to

The angle of contact.

*If the height of the meniscus is not wholly negligible in comparison with h, the mean value of h can be obtained by adding $\frac{1}{3}R$ to the height of the lowest point of the meniscus. For the volume of the liquid above this lowest point is the volume of a cylinder of radius R and height R, minus the volume of a hemisphere of radius R; or, $\pi R^3 - \frac{4}{6}\pi R^3 = \frac{1}{3}\pi R^3$. This volume divided by the area of the base, viz., πR^2, gives the mean height, viz., $\frac{R}{3}$. Formula (163) thus modified holds for tubes of as much as 2 mm. diameter.

the surface, the molecules would move over the surface in obedience to this component, i.e., equilibrium would not exist. If, then, *on* (Fig. 103) represent the line of junction of a liquid with a solid wall, f_1 the resultant of all the forces exerted upon the molecules at o by such portion of the liquid as lies within the molecular range when the liquid surface is assumed horizontal, and f_2 the resultant of the forces exerted upon the same molecules by the molecules of the wall which lie either above or below the horizontal line passing through o, then three cases may be distinguished:

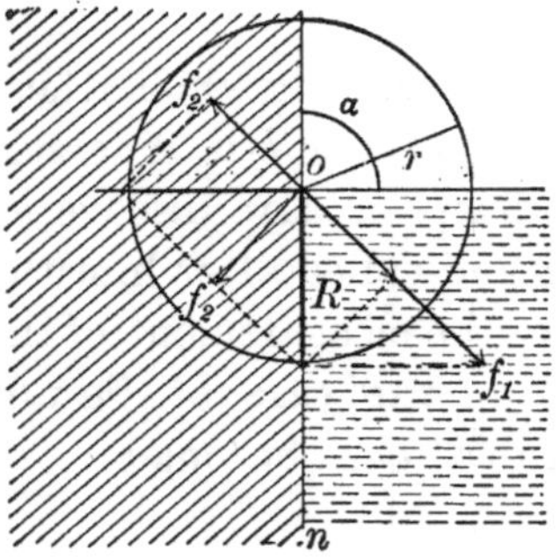

FIGURE 103

(1) That in which $f_1 = 2f_2$. In this case, as appears from Fig. 103, the cohesion of the liquid is exactly equal to twice the adhesion of the solid and liquid, and the final resultant R is parallel with the wall. Hence equilibrium exists in the condition assumed, i.e., the angle of contact α is 90°.

(2) That in which $f_1 > 2f_2$. The resultant R then falls to the right of *on*. Hence equilibrium can not exist until the surface near o has become convex, i.e., until the angle of contact α has become acute. This is the case of liquids which do not *wet* the wall. If the substances be mercury and glass (Fig. 104), equilibrium is reached when α is about 43°. It is to be observed, in general, that this angle α must always be the same for the same two substances. For, on account of the extreme minuteness of the sphere of influence, the weight of the particles contained within it is wholly negligible in comparison with the molecular forces, i.e., it is simply the relation between these molecular forces which determines the angle of contact.

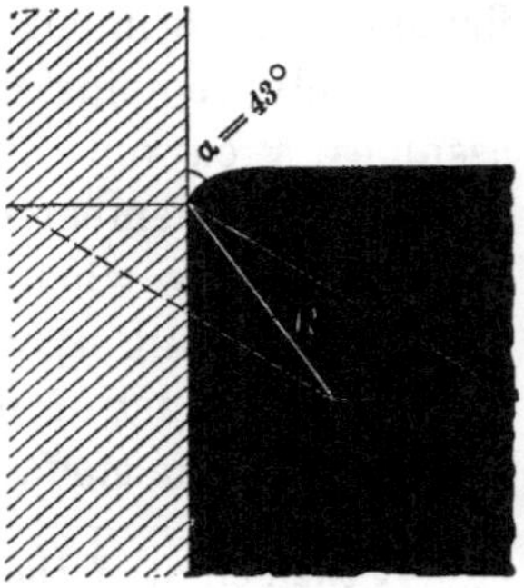

FIGURE 104

(3) That in which $f_1 < 2f_2$. The resultant R then falls to the left of *on* (see Fig. 105). Hence equilibrium can not exist until the surface near o has become concave and the angle of con-

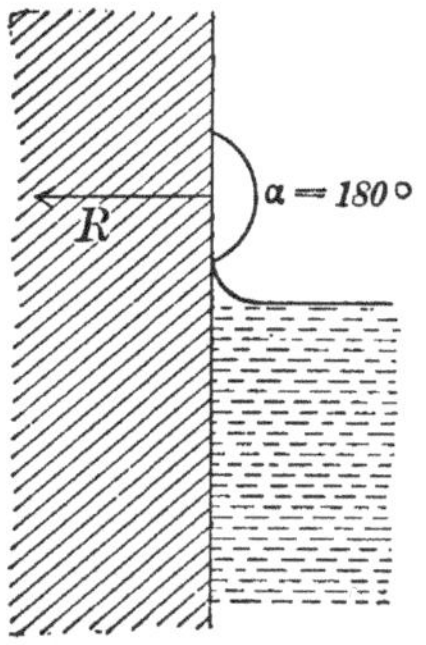

FIGURE 105

tact obtuse. In all cases in which the liquid completely *wets* the solid, the angle of contact is necessarily 180°, i.e., a thin film of the liquid lies flat up against the face of the solid. This is evident from the consideration that when a partially immersed body is raised from a liquid, the angle of contact can not remain constant at any value less than 180° unless the liquid retreats down the side of the body as rapidly as the body rises, i.e., unless the liquid be one which does not *wet* the solid.

The law of transmission of pressure by liquids easily accounts for this, at first view, somewhat surprising result. For, in accordance with this law, the molecular pressure P' existing because of adhesion at a point in the liquid close to the limit of contact (see Fig. 106), is transmitted undiminished in a direction parallel to the surface of the solid, and therefore constitutes a force pushing out the molecules at c. The only opposing force acting to prevent the limiting molecules from moving up along the surface is the vertical component of the attraction f exerted upon these molecules by such portion of the liquid as lies within the sphere of influence drawn about c. Hence, unless the ratio of the cohesion to the adhesion exceeds a certain limit, a thin film of the liquid must spread out indefinitely over the surface of the solid. This conclusion is not surprising, since it means simply that a body which attracts a liquid strongly enough will draw every particle of it as near as possible to itself.

Theory of thin liquid films on foreign surfaces.

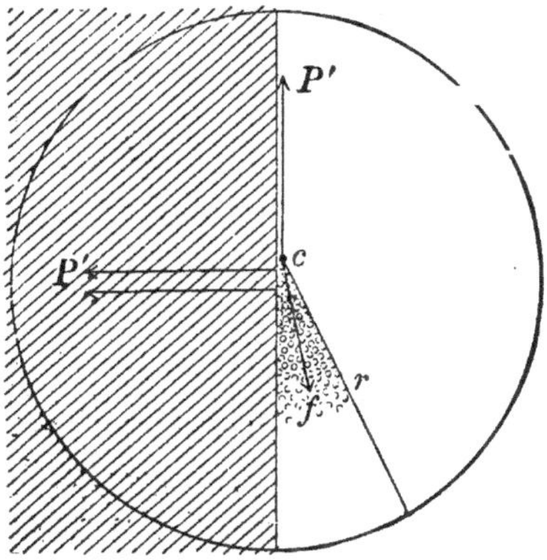

FIGURE 106

Thus it is that a drop of water spreads out indefinitely over a perfectly clean glass or mercury surface, that a drop of olive oil spreads over water, or, in general, that any liquid spreads out over any perfectly clean surface which it wets. But such perfectly clean surfaces are very difficult to obtain, and that on account of the prevalence of this very phenomenon. Thus, the least drop of oil touching a mercury or a glass surface spreads over it very quickly and completely changes the effect of adding a drop of water. However, such familiar facts as the creeping of salt solutions over battery jars, of kerosene over lamps, or the rapid spreading of oil over water, attest the correctness of the above conclusions. Of course, when but a drop of the liquid is present a limit to the spreading must be reached when the liquid attains a thickness of the order of magnitude of the diameter of the molecule. Rayleigh measured oil films on water which had a thickness of but .000002 mm. The diameter of an oil molecule can not, therefore, be more than this.

The theory of surface tension.

Another interesting result which may be deduced from Laplace's theory of molecular pressure is that, in general, a liquid must behave as though its surface were a stretched elastic membrane. For, since every molecule in the active layer is always being urged into the interior, it follows that as many molecules as can possibly do so will leave this layer and pass within, i.e., a liquid, like a distended rubber balloon, will always tend to draw together into the form which has the smallest possible surface for a given volume. Thus it is that all bodies of liquid which are not distorted by gravity or other outside forces always assume the spherical form, e.g., a raindrop, a soap bubble, a globule of oil floating beneath the surface of a liquid of the same density.

It follows again, from the tendency to assume the form of smallest surface, that a liquid film, a form of enormous surface, must exhibit a sensible *contractility*. Experiment amply supports the conclusion. Thus, a soap bubble may be observed to begin to draw back into the bowl of the pipe as soon as the blower removes his mouth.

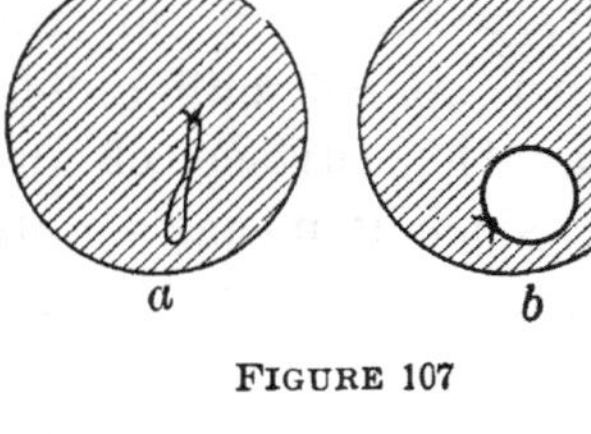

FIGURE 107

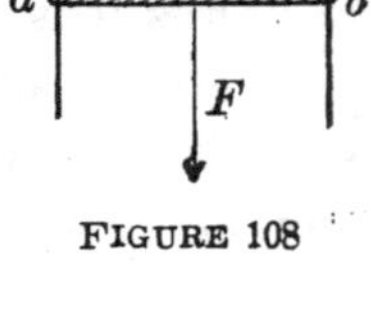

FIGURE 108

A wet loop of cotton thread laid upon a soap film, as in Fig. 107a, is snapped out at once into circular form, as in Fig. 107b, as soon as the film within the loop is pricked with a pin. A film formed in the frame *abdc* (Fig. 108) snaps the piece *ab* back to *cd* as soon as the stretching force F is removed.

Tension of films independent of thickness.

Further, Laplace's theory leads to the remarkable conclusion that the contractility of liquid films is wholly independent of their thickness. For the work which is performed by any agent which is increasing the surface of a liquid consists solely in bringing new molecules from the interior to the surface, against the force of the molecular pressure. Similarly the contractility of the film when the stretching force is removed is nothing but a manifestation of the force of molecular pressure drawing back molecules into the interior. Hence the work done by the outside agent when the surface is increasing, or by the molecular pressure when it is decreasing, is simply proportional to the increase or decrease in surface; i.e., the work required to pull down *ab* (Fig. 108) a given distance, e.g., 1 mm., must always be the same, whether the film has been stretched little or much, i.e., whether it is thick or thin. It is because this conclusion is at variance with the law which governs the stretching of solids (stretching force proportional to cross-section) that it appears strange. It is, however, completely confirmed by experiment. Thus the fact that the loop of Fig. 107b takes the accurately circular form shows that it is subjected to precisely the same force at all points on its circumference; yet the varying colors of the film show that it has a widely varying thickness.

It is to be observed, however, that this conclusion as to the constancy of F should hold only so long as the film is more than twice as thick as the active layer; for after this thickness has been reached the molecular pressure, and hence also the work required to bring a new molecule from the middle to the surface, must begin to diminish. It is probable, however, that the film must break at this point. Hence it is that the smallest thickness which a soap film can have is usually taken as a measure of the diameter of the sphere of molecular influence. This quantity, as measured by Johonnott at the Ryerson Laboratory in 1897, is .000012 mm.

It follows from the constancy of F that if *ab* (see Fig. 108) be pulled down a distance d, the work done by F is equal to Fd.

But this work is proportional to the increase in surface. If, then, T represent the amount of work which must be done against the molecular pressure in order to bring enough molecules into the active layer to form one new sq. cm. of surface, then, since the total increase in surface (considering both sides of the film) is $2ab \cdot d$, it follows that

Relation between molecular pressure and surface tension.

$$Fd = 2ab \cdot d \cdot T, \quad \text{or} \quad T = \frac{F}{2ab}. \qquad (164)$$

But $2ab$ is simply the length of the line of surface to which the stretching force F is applied. And since the value of F depends not at all upon the thickness of the liquid, but only upon the length of the attached surface line $2ab$, and upon a quantity T which is proportional to the normal molecular pressure, it is obvious that it is merely the surface of the liquid down to a depth r, i.e., the active layer, which is to be regarded as the seat of the contractile force F of the film. Finally then, since when $2ab = 1$,

$$T = F, \qquad (165)$$

it follows that *Laplace's molecular pressure manifests itself in any liquid surface as a tangential contractile force* (see Fig. 109) *which is numerically equal, in grams per cm. of length, to the quantity of work, expressed in gm. cm., which is required to bring up into the active layer, against the molecular pressure, enough molecules to form one new sq. cm. of surface.*

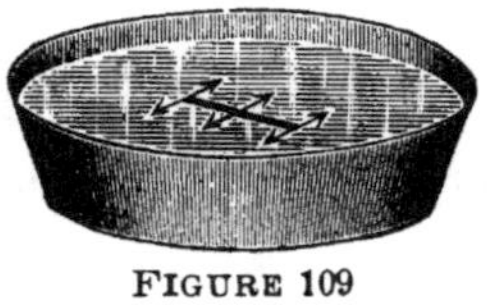

FIGURE 109

A rather interesting experiment has been devised to illustrate this fact of contractility when but *one* surface of a thick film is allowed to contract. A shallow vessel with one side cd movable about c is filled with water (see Fig. 110). As soon as the thread t is burned, the side cd is pulled over into the vessel, in spite of the fact that the weight of the liquid would tend to press it more firmly against the support e.

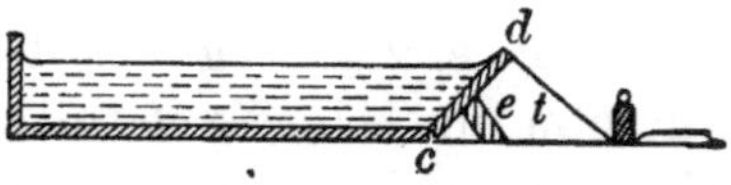

FIGURE 110

Now the pressure existing within a rubber balloon may be

easily calculated from a knowledge of the tension in its elastic envelope. Since, then, the molecular pressure in liquids manifests itself as a surface contractility, it ought to be possible to obtain, from the value of this contractility, the quantity p of (161), i.e., the increase in internal pressure which is due to a *curvature* of the surface. Thus let $\sigma\sigma'$ (Fig. 111) represent an infinitely small rectangular element of a convex surface. Let the arcs σ and σ' correspond to the two principal radii of curvature R and R'. If the tension in the surface has a force of T grams per unit length, then the number of grams of force F, which act on each of the arcs σ, is $T\sigma$. These forces are, of course, tangential to the surface and perpendicular to the arcs. Similarly, $T\sigma'$ grams of force act upon each of the arcs σ'. Since the surface is curved, all four of these forces have slight components in the direction of the normal *on*. The pressure beneath the element is evidently the sum of these normal components divided by the area $\sigma\sigma'$ of the element, for pressure is, by definition, force per unit area. The component of F parallel to *on* is $F \cos \alpha = F \sin \beta$. But in the limit

Deduction of Laplace's formula.

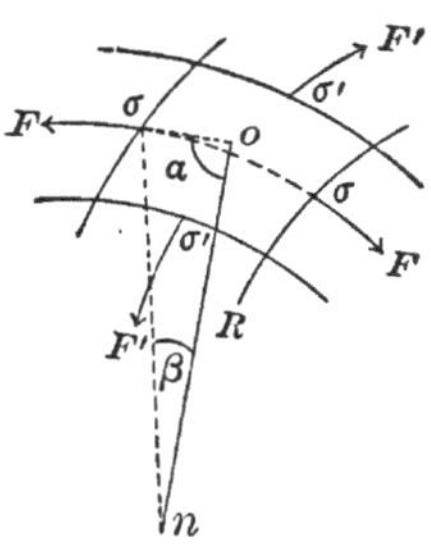

FIGURE 111

$sin\ \beta = \frac{\frac{1}{2}\sigma'}{R'}$. Hence the sum of the normal components of the two forces F is $\frac{T\sigma\sigma'}{R'}$. Similarly, the sum of the normal components of the two forces F' is $\frac{T\sigma\sigma'}{R}$. Hence the pressure p due to curvature is given by

$$p = \frac{\frac{T\sigma\sigma'}{R} + \frac{T\sigma\sigma'}{R'}}{\sigma\sigma'} = T\left(\frac{1}{R} + \frac{1}{R'}\right). \qquad (166)$$

That this formula is identical with that defined by Laplace by a more direct but much more difficult method attests the correctness of the reasoning, and also shows that *Laplace's capillary constant A is simply the tension in the surface per unit length or the amount of work required to add one sq. cm. to the surface.*

Experiment

Object. To compare the values of the surface tensions of water and alcohol, as given by the capillary tube method, with the results obtained by measuring directly the contractility of films.

DIRECTIONS.—**1.** Fill two small beakers, one with pure distilled water, the other with absolute alcohol. Then prepare a number of fresh capillary tubes by heating to softness bits of clean glass tubing in a Bunsen flame, and drawing them down to diameters of from .1 to .5 mm. Select several tubes which seem to be most nearly circular in form, and attach them by means of a rubber band to a mirror millimeter scale, as shown in Fig. 112. Take the reading r_0 of the fixed point *o* upon the scale, by placing the eye so that the image of *o* comes into coincidence with *o* itself.* Then immerse the lower ends of the tubes in the liquid and raise and lower the clamp *c* several times (a rack and pinion adjustment is to be preferred) so as to wet thoroughly the capillaries above the points reached by the liquid. Next bring *o* exactly into contact, from below, with the liquid surface, slip up the capillary alone a trifle, and take the reading *r* of the bottom of the meniscus as soon as the level has settled back to its position of equilibrium. It is clear that the height of rise *h* is given by $h = r - r_0$. Mark by means of a bit of wax the point to which the liquid rises in the tube, then remove it from the scale, scratch it very carefully with a sharp file at this point, and break it off as squarely as possible. Stick the broken tube upright against the side of a block of wood by means of soft wax. Then focus upon the broken end a microscope which is provided with a micrometer eyepiece. Count the number of turns† and fractions of a turn

The height of rise h.

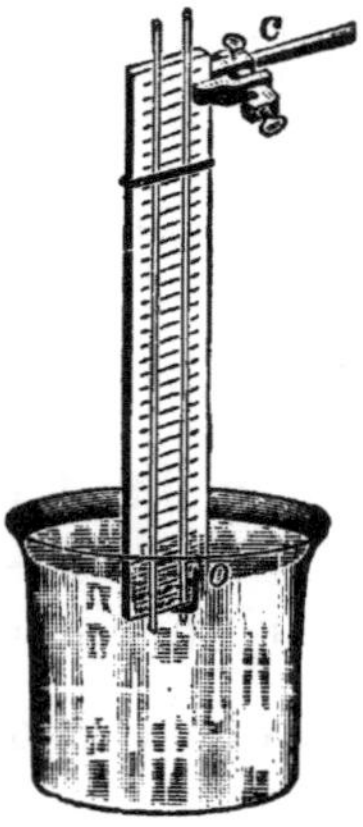

FIGURE 112

The radius R of the tube.

*If a mirror-scale is not available the eye may be set in the correct position for reading upon an ordinary scale, by pressing a small piece of mirror glass against the scale, behind the point *o*.

†The counting is greatly facilitated by means of the toothed edge which is found in the field of view of the eyepiece. Each tooth corresponds to one revolution.

which must be given to the micrometer screw in order to cause the movable cross-hairs to traverse exactly the internal diameter of the tube. Repeat several times, using in each case a different diameter. Then replace the capillary tube by a standard millimeter scale, and find in the same way the number of turns and fractions of a turn corresponding to 1 mm. From the two observations find the diameter D of the tube in mm.

In (163) hd represents a pressure expressed in grams per square centimeter. Hence, in order to obtain A in absolute units, the hd of (163) must be multiplied by 980, and both R and h must be expressed in centimeters. The best determinations have given for water at 15°, $A = 75$; for alcohol, $A = 25.5$.

Calculation.

If the results obtained by this method are not uniform, it will be because, on account of the presence of impurities, the wetting of the tube is not perfect, or because the tube has not a circular section. It is to be observed also that A is a function of the temperature, diminishing as the latter increases. This was to have been expected, since a rise in temperature corresponds to a pushing apart of the molecules.

Sources of error.

2. In order to make a *direct* measurement of the surface tension, attach a very light wire frame a (Fig. 113) to a delicate helical spring s, and by means of an elevating table b, raise a vessel of liquid till the frame is immersed. Next lower the table carefully by means of a rack and pinion r, until a film forms between the prongs of the frame. Then quickly take the reading of the index i upon the mirror-scale m. Before repeating, stir the liquid vigorously by means of a glass rod which has been carefully cleaned in a Bunsen flame. Continue this operation until a number of consistent readings can be obtained. The difference between this reading and that taken when the spring and frame hang freely, is, of course, a measure of the force of tension F possessed by the two surfaces of the film. In order to reduce this force to dynes observe the elongation produced by a known weight of the same order of magnitude as

The direct measurement of T.

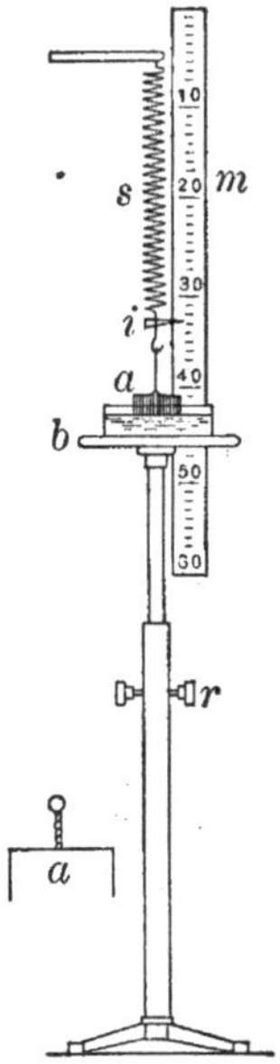

FIGURE 113

F. Then apply Hook's Law to determine F accurately in grams. Finally measure the distance l between the vertical wires of the frame a with an ordinary metric scale and calculate T from (164).

Since the presence of the least particle of oil upon the surface changes completely the value of T, it is of great importance that the frame and vessel be thoroughly cleaned with caustic potash before the experiment is begun, and that care be taken not to touch the liquid at any time with the fingers. The purpose of the stirring after each observation is to break up any film of impurity which may be present in spite of all precautions. It will usually be found to increase the reading somewhat. The readings are to be taken only when a distinct film is visible between the prongs. If the frame continually snaps up without forming a film, clean it again with caustic potash and lower the table more slowly.

Precautions.

Record

1. $r_0 =$ ——	No. turns of microscope micrometer to one mm. = ——		
Water		Alcohol	Density = ——
r_1 (tube 1) = ——	r_2 (tube 2) = ——	r_3 (tube 3) = ——	r_4 (tube 4) = ——
$\therefore h_1 =$ ——	$h_2 =$ ——	$\therefore h_3 =$ ——	$h_4 =$ ——
D_1 (in screw turns) = ——	$D_2 =$ ——	$D_3 =$ ——	$D_4 =$ ——
$\therefore A_1 =$ ——	$A_2 =$ ——	$A_3 =$ ——	$A_4 =$ ——
	Mean A $[= T] =$ ——		Mean A $[= T] =$ ——
2. Rd'g with film = ——	zero = ——	Rd'g with film = ——	zero = ——
" " —— mg = ——	" = ——	" " —— mg = ——	" = ——
$\therefore F =$ —— $l =$ ——	$\therefore T =$ ——	$F =$ —— $l =$ ——	$\therefore T =$ ——

Problems

1. The gifted American physicist, Joseph Henry, first suggested in 1848 the determination of the capillary constant by attaching a manometer to a soap bubble, and thus measuring the pressure existing within the bubble. Assuming the surface tension of a soap solution to be 70 dynes, find what would be the difference in the levels in the arms of a water manometer when attached to a bubble of 5 cm. diameter.

2. Find how many ergs of work must be expended to blow a soap bubble of 15 cm. diameter.

3. A drop of water placed in a conical tube (see Fig. 114) is observed to travel rapidly toward the small end; but a drop of mercury travels toward the large end. Explain

FIGURE 114

4. How high will water rise in pores which are .001 mm. in diameter?

5. Deduce formula (163) from the consideration that, in a capillary tube in which the angle of contact is 180°, the total upward force is the surface tension acting upon a line whose length is the circumference of the tube, while the balancing downward force is the weight of the liquid raised.

6. Explain from considerations of molecular pressure, how a needle or any small body which is much more dense than water, may yet float upon it provided $\alpha < 90°$. Could it ever float if $\alpha > 90°$?

XXII

CALORIMETRY

Theory

The nature of heat. The definition of the calorie.

Calorimetry is that branch of Physics which deals with the measurement of *heat quantity*. It had its beginning about the year 1760 with the work of the Scotch chemist and physicist, Joseph Black, the originator of the caloric theory, and the first investigator to draw a sharp distinction between *heat* and *temperature;* in fact the first to give any careful definition of the term heat.

That bodies change in temperature is a fact of direct observation, but the notion that a something called heat passes between bodies of changing temperature is of the nature of an hypothesis. This hypothesis has taken two forms. With Black and his followers, the so-called calorists, heat was an imponderable fluid, *the caloric*, the passing of which into or out of a body was the cause of temperature change. The *unit of heat*, *the calorie*, was then defined as *the amount of heat which must enter or leave 1 gram of water in order to produce 1 degree of change in its temperature.*

With Joule, Clausius, and practically all physicists of the latter half of the nineteenth century, a rise in temperature represents an increase, not in the quantity of a contained heat fluid, but simply in the mean kinetic energy of the molecules. The calorists' definition of the heat unit has, however, been retained in its original form, their concept of the transfer of a heat fluid being simply replaced by the concept of a transfer of molecular energy, kinetic or potential, or both. A knowledge of the caloric theory is now important only because of the light which it throws upon the terminology of heat. The theory was altogether abandoned after Joule's demonstration of the equivalence of heat and work.

Up to Black's time it was generally supposed that the rise in temperature of a substance in contact with a hot body was continuous; but Black pointed out that while ice or snow is changing into water it maintains, if well stirred, a perfectly constant temperature, no matter how hot may be the stove with which it is in contact. In order to explain this phenomenon, together with the inverse one that the condensation of steam or the freezing of water is accompanied by a large evolution of heat, Black assumed that a certain amount of the caloric always became *hidden* or *latent* at the time of a change from the solid to the liquid, or from the liquid to the gaseous condition. For example, since it was found that the mixing of 1 gram of ice at 0° and 1 gram of water at 80° C. would yield 2 grams of water at 0°, or that 2 grams of water at 40° was required to just melt 1 gram of ice at 0°, 80 calories was taken to be the *latent heat of fusion of ice.*

Origin of the term "latent heat."

According to the modern mechanical theory, the temperature of a substance remains constant while the change of state is going on simply because the energy of motion communicated to the molecules in contact with the hot body is at once transformed into energy of position; that is, the heated molecules immediately break away from the forces which have been holding them in the given state (solid or liquid, as the case may be), and thereby lose their increased velocities (i.e., their increased temperature) as rapidly as they receive them. The operation is wholly analogous to that in which a body shot up from the earth loses its velocity in raising itself against gravity. Thus, although the old terms of the calorists, *latent heat of fusion* and *latent heat of vaporization*, are still retained, these latent heats represent to-day only given changes in the potential energy of the molecules, just as a given rise in temperature represents a given change in their mean kinetic energy.

Significance of latent heats.

But it must not be supposed that changes in the kinetic and potential energies of the molecules may not take place simultaneously. In fact, there is but a limited number of substances, those in general which are of a crystalline structure, which show at any points a change in potential energy unaccompanied by a change in kinetic, i.e., by a rise in temperature. Thus wax, resin, gutta percha, glass, alcohol, carbon, and a great number of other substances pass gradually through all stages of viscosity in melting or solidifying. In such cases the temperature changes *continually;* i.e., there is no definite point at which the substance may be said to begin to melt. On the other hand, *every* increase in the *temperature* of a solid is accompanied by a certain amount of expansion, and hence by some increase in the potential as well as the kinetic energy of the molecules.

The following table shows the values of the latent heats of some of the commoner substances:

	Melting point, °C.	Latent heat of fusion (calories)	Boiling point, °C.	Latent heat of vaporization (calories)
Water	0.	79.9	100.	536.
Benzol	5.3	30.	80.2	94.
Acetic acid	16.5	46.	118.	97.
Mercury	-39.5	2.8	357.	62.
Sulphur	114.	9.	447.	362.
Silver	970.	21.	——	——
Lead	328.	6	——	——
Platinum	1780.	27.	——	——

It was discovered very early that the quantity of heat given up by 1 gram of water in falling through 1 degree would raise very different weights of other substances through one degree, e.g., 33 gm. of mercury, 10.5 gm. of copper, 8.9 gm. of iron, 2.3 gm. of turpentine, etc. The calorists explained these facts by the assumption that different substances possess per unit weight different *capacities* for the heat fluid. *The heat capacity of a body* was then defined as *the number of calories required to raise the body through 1 degree*, and *the specific heat of a substance*, as *the number of calories required to raise 1 gram of that substance through 1 degree.* These definitions are still retained now that heat is regarded as molecular energy; but the fact that different amounts of this energy must be communicated to gram weights of different substances in order to produce the same increase in temperature, i.e., the same increase in the average kinetic energy of translation of the molecules, is attributed to

Specific heat.

(1) The differences in the number of molecules contained in gram weights of different substances; and

(2) The differences in the *internal work* which are incidental to an increase in temperature. By internal work is meant the work done in increasing the distances between the molecules, in augmenting the energy, kinetic or potential, of the atoms within

the molecules, or, in general, any work other than that represented in the increase in the kinetic energy of translation of the molecules themselves, or in the expansion against the force of atmospheric pressure.

Molecular heats.

The first element can be easily investigated, for if this were the only cause of difference in the specific heats of different substances, these differences would disappear in a comparison of quantities which represent, not equal weights, but equal numbers of molecules. Such quantities can evidently be obtained by taking, in each case, a number of grams which is equal to the molecular weight of the substance. This quantity has been given the name of a *gram-molecule*, and *the number of calories of heat required to raise 1 gram-molecule of a substance through 1° C. is called its molecular specific heat*, or simply *its molecular heat*. The molecular heat is evidently, therefore, simply the product of the specific heat per gram and the molecular weight. The following table contains a comparison of a very few specific heats per given weight, and specific heats per given number of molecules:

	Gases	Specific Heat	Molecular Weight	Molecular Heat
1	Oxygen (O_2)	.2175	32.	6.95
	Nitrogen (N_2)	.2435	28.	6.81
	Hydrogen (H_2)	3.4090	2.	6.82
	Hydrochloric acid (HCl)	.1845	36.4	6.72
	Carbon monoxide (CO)	.2450	28.	6.86
	Nitric oxide (NO)	.2317	30.	6.95
2	Nitrous oxide (N_2O)	.2262	44.	9.95
	Carbon dioxide (CO_2)	.2163	44.	9.52
	Water vapor (H_2O)	.4805	18.	8.64
	Solids			
3	Potassium (K_2)	.1655	78.2	12.94
	Sodium (Na_2)	.2934	46.	13.50
	Silver (Ag_2)	.0570	216.	12.32
	Copper (Cu_2)	.0952	126.8	12.08
	Mercury (Hg_2)	.0332	400.	13.28
4	Oxide of copper (CuO)	.1420	79.4	11.27
	Oxide of nickel (NiO)	.1588	74.8	11.87
	Oxide of mercury (HgO)	.0518	216.	11.19

	Solids	Specific Heat	Molecular Weight	Molecular Heat
5	Chloride of calcium ($CaCl_2$)	.1642	111.	18.2
	Chloride of zinc ($ZnCl_2$)	.1362	136.2	18.6
	Chloride of barium ($BaCl_2$)	.0896	208.	18.6
6	Sulphate of lead ($PbSO_4$)	.0872	303.	26.4
	Sulphate of barium ($BaSO_4$)	.1128	233.	26.3
	Sulphate of calcium ($CaSO_4$)	.1966	130.	26.7

This table shows that many of the differences in specific heats do in fact disappear upon comparison of equal numbers of molecules. The differences which are still left must be attributed wholly to the second cause, viz., differences in internal* work because of differences in molecular structure or molecular attraction. Since an extensive series of observations, of which the above table is a small fraction, has shown that, in general, the molecular heats of chemically similar substances, the molecules of which possess the same number of atoms (see table), are nearly the same in a given state of aggregation (law of Neumann), it must of course be inferred that the internal works are also the same for such similar substances.

Since the molecules of gases are not subjected to appreciable mutual attractions it might be expected that with molecules of equal complexity the molecular work would be less in the gaseous than in the solid or liquid condition (see groups 1 and 3). Again, it would be natural to conclude that, for substances in the same state of aggregation, the internal work would in general increase with the complexity of the molecule. Both of these inferences are seen to be in accordance with the facts presented in the table.

The law of Neumann† is not exact, nor could it be expected to

* The external work done in expanding against atmospheric pressure may be neglected for solids and liquids. For gases it is a constant quantity.

† The law of Neumann is an extension of a law discovered in 1818 by Dulong and Petit in accordance with which the atomic heats (products of specific heat and atomic weight) of all the solid elements are nearly the same, amounting to about 6.3 (see group 3 of table). This law was extended in another direction in 1848 by Woestyn who found that the molecular heat of a compound is often equal to the sum of the atomic heats of the elements contained in the compound. It is difficult to recognize in this discovery anything more than an interesting empirical law.

Atomic heats.

be in view of the differences in the attractions which exist between different sorts of molecules. In fact, experiment has shown that the specific heat of a given substance is not constant, but that it in general increases steadily with the temperature. Hence, save in the case of the permanent gases, the quantities given in the table are only to be regarded as *mean* specific heats between certain temperatures, e.g., 15° and 100°. The rate of increase is fortunately slight for water and for most solids, so that in ordinary work at moderate temperatures, it may be disregarded. But for most liquids it is far from negligible; for example, the specific heat of alcohol is .54 at 0° and .64 at 40°C. The fact of the dependence of the specific heat of water upon the temperature was first established by Regnault by the method of mixture. Thus, for example, it was found that equal quantities of water at different temperatures do not, when mixed, yield exactly the mean temperature; or, again, that the fall of 10 gm. of water from 60° to 30° does not heat a given quantity of cold water quite as much as the fall of 5 gm. from 90° to 30°. The first exact work upon the nature of this variation was, however, done by Professor Rowland of Johns Hopkins in 1879, in accordance with whose results the specific heat of water diminishes by about 1 per cent from 0° to 29°, and then increases again slowly to 100°.

Variation of specific heats with the temperature.

In view of these facts it has been found necessary *to define the calorie as the amount of heat required to raise 1 gm. of water, not through 1 degree, but from 15° to 16° C.* The definitions of heat capacity and of specific heat remain unchanged, but the quantity usually obtained by experiment is not the specific heat at a given temperature, but rather the mean specific heat between two specified temperatures. Thus, if Q represent the amount of heat passing into or out of a mass m while it changes in temperature from t_1 to t_2, then

Modified definitions.

$$\frac{Q}{t_2 - t_1} = \text{mean heat capacity between } t_1 \text{ and } t_2; \qquad (167)$$

$$\text{and } \frac{Q}{m\,(t_2 - t_1)} = \text{mean specific heat between } t_1 \text{ and } t_2. \qquad (168)$$

Measurement of heat.

There are three methods which have been used for the measurement of specific and latent heats. These are:

(1) The method of mixture,

(2) The method of cooling,

(3) The method of fusion of ice or condensation of steam.

The method of mixture.

The method of mixture is most common, most simple, and, in many cases, most accurate. It consists in mixing known weights of bodies of different temperatures, observing the resulting temperature and then writing out an equation which contains *on one side all of the heat quantities lost by the cooling bodies, and on the other, all of the heat quantities gained by the warming bodies.* The specific heat, or the latent heat, sought is the unknown quantity of this equation. For example, suppose that it be required to find the latent heat of steam x from an experiment in which p gm. of steam at 100° are condensed in M gm. of water. Let t_i be the initial temperature of the water and t_f the final temperature attained by the mixture. Then the number of calories lost by the steam in condensing is px; that lost by the condensed steam, in passing from its temperature of condensation, 100°, down to t_f, is $p(100 - t_f)$. The heat gained by the water is $M(t_f - t_i)$. Hence, if no heat went into the vessel, the thermometer, the stirrer, or the atmosphere, the equating of the heat losses and heat gains would give

$$px + p(100 - t_f) = M(t_f - t_i). \qquad (169)$$

Calorimeter and thermometer corrections.

But, as a matter of fact, the calorimeter gains heat as well as the water. If its heat capacity be represented by C, the number of calories so gained is $C(t_f - t_i)$. Similarly, if C' and C'' are the heat capacities of the thermometer and stirrer respectively, $(C' + C'')(t_f - t_i)$ calories go into them. These terms must therefore be added to the right side of (169). The quantities C, C', and C'' are determined either from direct observation (see Experiment) or from the product of the known weights and specific heats.

Radiation correction.

To determine the number of calories gained by the atmosphere through radiation from the calorimeter it is necessary to know (1) the mean temperature t_m of the water during the experiment, (2) the mean temperature t_j of the surrounding atmosphere, (3) the number of minutes of duration N of the experiment, and (4) a quantity k called the *radiation constant* of the calorimeter. This constant represents *the number of calories which will pass per minute into the atmosphere from the calorimeter when the temperature of the latter is 1° above that of the surrounding air.* Then in the above experiment the number of calories L lost by radiation is

$$L = k\ (t_m - t_j) N. \tag{170}$$

This also must be added to the right side of (169). Its sign will of course be negative if $t_m < t_j$.

t_m is determined by averaging the temperature readings made at intervals of 15 seconds throughout the experiment. k is obtained from an observation of the number of minutes n required for the temperature of the water to fall, e.g., .5°, after the conclusion of the experiment. If, then, d be the mean difference in temperature between the water and the room during this time of fall, since the total number of calories lost during the n minutes is $M \times .5$, the number of calories which would be lost per minute for a difference of temperature of 1° is

$$k = \frac{M \times .5}{dn}. \tag{171}$$

It will be observed that in this method of calculation it is assumed that a body radiates heat, i.e., falls in temperature, at a rate which is proportional to the difference between its temperature and that of the surrounding atmosphere. This is a law of cooling which was announced by Newton,—a law which is *not even approximately correct.* However, where the radiation correction is small and the quantity $(t_m - t_j)$ not more than 5°, the incorrectness of Newton's law will not generally introduce an appreciable error into the result.

Precautions for reducing the radiation correction.

The radiation correction, however, can never be determined with great certainty. Hence the effort is always made to reduce it as much as possible. This is done (1) by using such large quantities of water that the temperature change is small, (2) by making the initial temperature about as much below the room temperature as the final is above it, (3) by highly polishing the outside surface of the calorimeter so that it radiates very slowly, and (4) by inclosing it in a still air space surrounded by constant temperature walls. With the form of calorimeter shown in Fig. 115, the radiation correction is ordinarily negligible unless the experiment lasts more than

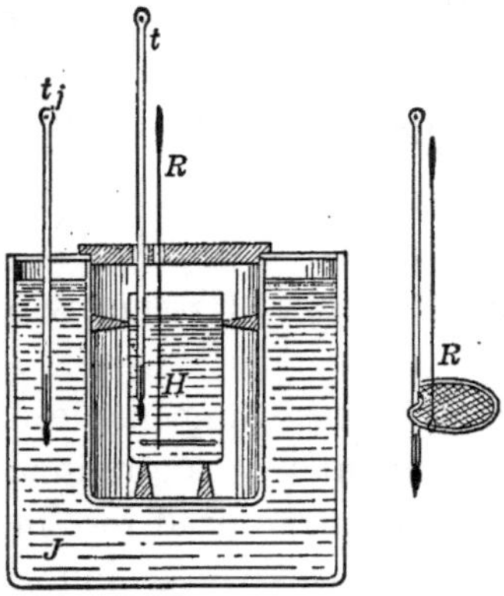

FIGURE 115

2 minutes or unless the difference of temperature between the calorimeter H and the water-jacket J exceeds 2° C. A very sensitive thermometer is of course necessary for the accurate measurement of such small temperature changes.

The temperature of the hot body.

The above discussion applies in general to the *method of mixture*, whether it be a latent or a specific heat which is sought. But in the latter case some special device must often be employed for obtaining accurately the temperature of the hot body, and for transferring it quickly to the calorimeter. Solids which are not acted upon by water are placed in finely divided form, in a wire net, and heated to a temperature indicated by a thermometer b in an inclined air tube F which is surrounded by a water or steam bath P (see Fig. 116). In cases in which there might be any heat-producing action between the water and the solid, the latter is first inclosed in some thin-walled vessel of known heat capacity, then heated in F and dropped into H as before. In the case of liquids, no special heating device is necessary, since thermometers can be plunged directly into them; but such liquids as can not be brought into contact with water must be inclosed, like similar solids, in thin-walled vessels of known heat capacity.

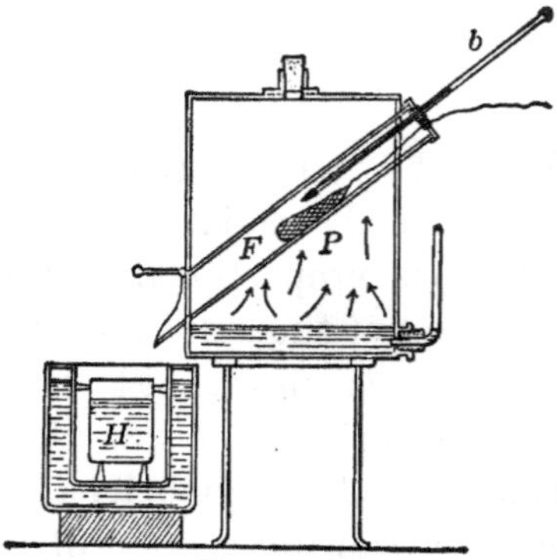

FIGURE 116

The method of cooling.

The second method of determining specific heats is the *method of cooling*. It consists in comparing the times required for a given closed vessel to cool, in air, through a given number of degrees, first when filled with water and then when filled with the substance whose specific heat is sought. If Z_1 and Z_2 are the two times of cooling, and C_1 and C_2 the two corresponding heat capacities of the vessel and contents in the two cases, then it may be shown that

$$\frac{Z_1}{Z_2} = \frac{C_1}{C_2}. \qquad (172)$$

For, since radiation takes place from the surface layers only, it is clear that, with given outside conditions, a given surface, at a given temperature, must always lose heat at the same rate, no matter what substance may be inclosed by the walls. Let, then, q denote the number of calories which passes out of a given surface at temperature t in an infinitely short element of time. This loss in heat will be associated with a small fall in temperature, viz. δ_1, which will be determined by [see (167)]

$$q = C_1\delta_1. \qquad (173)$$

If now the heat capacity be changed from C_1 to C_2 by a change in the contents of the vessel, the new fall in temperature δ_2 in the same infinitely short interval of time and at the same temperature t will be determined by

$$q = C_2\delta_2. \qquad (174)$$

Hence, from (173) and (174),

$$\frac{\delta_1}{\delta_2} = \frac{C_2}{C_1}, \qquad (175)$$

i.e., at any given temperature of the surface the two *changes in temperature* during the same small interval *of time* are *inversely* proportional to the heat capacities. This is exactly equivalent to the statement that, at a given temperature, the two *intervals of time* required for the same small change in *temperature* are *directly* proportional to the heat capacities. Since, then, the times required to pass through each small element of the scale, e.g., from 60° to 59°, are proportional to the heat capacities, the *total* times required to pass through any interval of temperature made up of these small elements must also be proportional to the heat capacities.

It is to be observed that this conclusion involves no assumption whatever regarding the nature of the law of cooling, or regarding the relation between the rôles played by convection currents and by true radiation in the cooling process. It rests solely upon the assumption of similarity in the outside temperature conditions and uniformity of temperature in all parts of the cooling body. This last condition is

Conditions in which method of cooling is applicable

difficult to fulfil when the cooling vessel contains solids. Hence the method has not proved satisfactory for the determination of the specific heats of such substances; but for liquids it has been found both accurate and convenient.

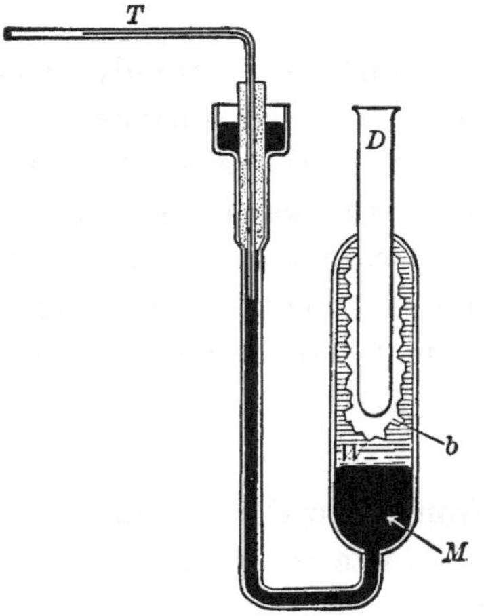

FIGURE 117

The ice and steam calorimeters.

The third method of measuring specific or latent heats consists in either determining the amount of ice m which a heated body will melt while its temperature is falling to 0°C., or finding the amount of steam m' which a cold body will condense while its temperature is rising to 100°C. The heat given up by the body in the first case is then evidently $80m$. That taken up in the second case is $536m'$.

The ice calorimeter is as old as Black, but the modern form is due to Bunsen (see Fig. 117). The ice cap b is formed in the water W by inserting a freezing mixture into the tube D. The point to which the mercury M rises in the graduated capillary tube T is then noted. The hot substance is next dropped into D, where it melts a certain amount of ice. The movement of the mercury in T to the right because of the contraction due to change of state is proportional to the amount of ice melted. The value in calories of one division of T is determined by inserting into D a substance of known heat capacity. This calorimeter has proved very valuable in determining the specific heat of very small bodies.

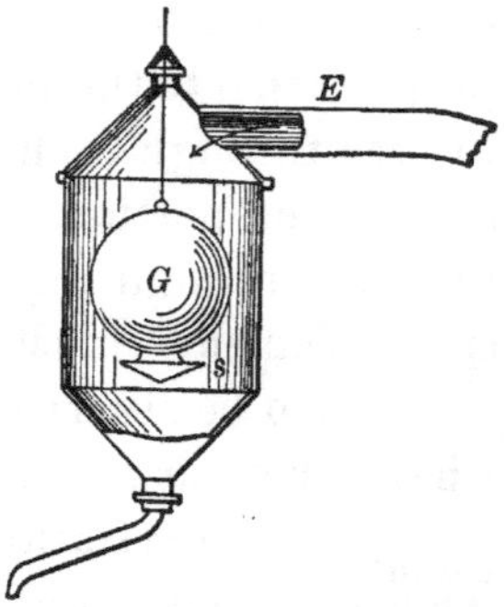

FIGURE 118

The steam calorimeter (Fig. 118) has proved of especial value only in the determination of the specific heats of gases. A light metal globe G, full of the gas, is suspended from the arm of a balance within a chamber of known temperature t. When steam is suddenly admitted into this chamber through E it condenses upon the globe and the walls until their temperature reaches

100°. With proper precautions to prevent the loss of drops of condensed steam (see pan *s*), the increase m' in the weight of the globe represents the amount of steam which must be condensed in order to raise the temperature of globe and contained gas from $t°$ to 100°. If the heat capacity C of the globe is known, that of the contained gas C' can evidently be found from $(C + C')(100 - t) = 536m'$. Accurate results can not be obtained with either of these latent heat calorimeters without the use of greater precautions than can be taken ordinarily in intermediate laboratory courses.

Experiment

Object. To compare the method of cooling and the method of mixture in the determination of the specific heat of turpentine.

DIRECTIONS.—**1.** By means of the trip scales find the weight w_c of the nickel-plated brass vessel A of about 30 cc. capacity (see Fig. 119). Then fill it with boiling water and weigh again (w_h). Subtract and obtain the weight of the water w_w. Treat the blackened vessel B in the same way, filling it with the same number of grams of water. Suspend both vessels from the wooden cover D by means of corks and thermometers, as shown in the figure. Attach the cylindrical brass vessels E and F by means of the catches at e, and immerse in a large pail or battery jar full of water at about the room temperature.

Observations upon rates of cooling.

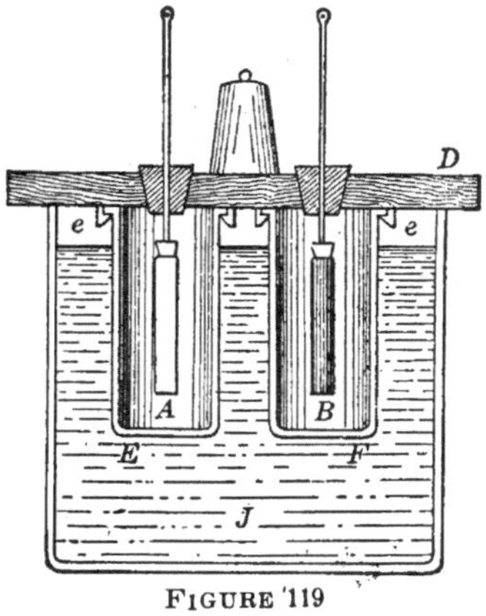

FIGURE 119

Take a very exact reading of each thermometer about once a minute while the temperatures are falling from 71° to 39°C. The hour, minute, and second of each reading must be accurately taken. Since the eye can not be upon both the watch and the thermometer at the same time, it is advisable to have in the room some audible second marker with which the watch can be compared just before each reading. This may be dispensed with if there are two observers, one to record temperatures and the other to record times. In this case the first observer should tap sharply upon the table at the instant of each temperature reading, at the same time calling out either *A* or *B*, while the second observer records in the *A* or *B* column the hour, minute, and second of each tap. The mean temperature t_j of the water-jacket may be found from observations taken at the beginning and end of the cooling process.

Next pour out the water from *A* and *B* and dry them thoroughly; fill to the same level as before, this time with hot turpentine; obtain the weights (w'_b), and by subtraction the corresponding weights of the turpentine w_t; replace in the bath and take a new set of observations upon the rate of cooling between 71° and 39°.

Plotting curves of cooling.

With the observed values of the times and temperatures, plot upon a large sheet of coördinate paper four smooth full-page curves, using times as abscissae and temperatures as ordinates. In so doing, choose the scale of temperature so that the lowest observed temperature is represented by a line near the bottom of the page, the highest by a line near the top. Choose the scale of times so that the time of beginning of observations upon the cooling of the polished vessel when filled with water is represented by a line (the zero of time for this case) which is near the left side of the page, while the time of conclusion of observations upon this case is represented by a line near the right side of the page. Plot the other curves upon the same sheet to the same scale, the zero of times being in each case the time of beginning of observations.

Computation.

Now read off very carefully upon the four smooth curves the four times included between any two temperature lines, e.g., 70° and 40°, and thus obtain from both *A* and *B* the quantity $\frac{Z_1}{Z_2}$. If, then, c and c' represent the respective heat capacities of the empty vessel *A* and of the thermometer, and s, the specific heat of turpentine, the equation $\frac{Z_1}{Z_2} = \frac{C_1}{C_2}$ becomes

$$\frac{Z_1}{Z_2} = \frac{w_w + c + c'}{w_t s + c + c'}. \qquad (176)$$

From (176) s can easily be obtained as soon as c and c' have been found.

To obtain c, multiply w_c by the specific heat of brass (.095). To obtain c', fill vessel A with water at the room temperature, immerse and read carefully the thermometer; then plunge it into hot water, withdraw, wipe off the adhering water; read quickly and again instantly immerse in A. Stir and note the rise in the temperature of the water. If then w' is the weight of the water, t_i and t_f its initial and final temperatures, and t the initial temperature of the thermometer, then [see (167)]

Heat capacities of vessels and thermometers.

$$c' = \frac{w' (t_f - t_i)}{t - t_f}. \tag{177}$$

From any two of the four cooling curves test the incorrectness of Newton's law of cooling as follows: If z_1 and z_2 are the times required to cool from 71° to 69° and from 41° to 39° respectively, and if t_j is the temperature of the jacket, then if Newton's law were correct it would follow that

Newton's law of cooling.

$$\frac{z_1}{z_2} = \frac{40 - t_j}{70 - t_j}. \tag{178}$$

2. To find the mean specific heat of turpentine between 70° and 40°C. by the method of mixture, first heat the water in the jacket J (Fig. 115) to about 37° and hold it constant at that temperature by the gentle application of heat and by such stirring as is found necessary. Find the weight W_c of the nickel-plated calorimeter H, including the stirrer, then fill half-full of water and weigh again (W_b). Represent the weight of the water alone by W_w. Give the water a temperature which is about 3° below that of the jacket and set it in place within the jacket as in Fig. 115. From the specific heat found by the method of cooling, calculate about how many grams of turpentine will need to fall from 70° to 40° in order to raise the water in the calorimeter from 34° to 40°. From this and the density of turpentine (.87) estimate very roughly to what height the mixture of water and turpentine will raise the level of the liquid in the calorimeter. Then heat to about 73° a half-liter or more* of turpentine in a dipper provided with a lip; stir it thoroughly with a thermometer (No. 1) until the temperature falls to about 70°, then take an accurate reading t'_i and very quickly pour

The method of mixture.

* A large quantity is used so that the cooling may not be too rapid.

about the estimated volume into the water. The temperature t_i of this water should have been taken but an instant before with the aid of a more sensitive thermometer (No. 2).

Cover the calorimeter as quickly as possible after the mixing and stir very thoroughly with the wire-net stirrer R (see Fig. 115), keeping it, however, always below the surface. Take a reading every 15 seconds from the time of mixing until the temperature has passed its highest point and begun to fall. The first few of these readings will of course be uncertain, but they have a very small influence upon the result. Record the highest temperature reached t_f, then withdraw the thermometer, shake back into the calorimeter the drops which adhere, and take the weight W_m of the calorimeter and contents, including stirrer. By subtraction obtain the weight W_t of the turpentine.

Correction for radiation.

To find the radiation constant pour out the mixture and fill the calorimeter with a volume of water about equal to the volume of the mixture. Find the weight M of the water alone, then raise it to about the final temperature t_f, replace it in the jacket and note the time n required, with continual stirring, to fall .5°C. From this and the temperature of the jacket, compute first the radiation constant (equation 171), then the number of calories lost by radiation (equation 170).

Corrections for calorimeter and thermometer.

Calculate the heat capacities of the calorimeter and stirrer by multiplying their joint weight by .095, obtain the heat capacity of the thermometer either by the method employed in 1 or by the following process: Estimate the volume in cubic centimeters of that portion of the thermometer which is immersed, by noting the rise of water in a narrow graduate when the thermometer is sunk in it up to the point to which it was wet in the experiment. Multiply this by the specific heat of mercury per cubic centimeter, viz. $13.6 \times .033 = .45$. It is because this is about the same as the specific heat of glass per cubic centimeter, viz. $2.5 \times .19$ that the thermometer may be treated as though it were made entirely of mercury.

Record

1.

A		B	
$w_c =$ ——		$w_c =$ ——	
$w_b =$ ——	$w'_b =$ ——	$w_b =$ ——	$w'_b =$ ——
$\therefore w_w =$ ——	$\therefore w_t =$ ——	$\therefore w_w =$ ——	$\therefore w_t =$ ——
$Z_1 =$ ——	$Z_2 =$ ——	$Z_1 =$ ——	$Z_2 =$ ——
$t_j =$ ——	$t_j =$ ——	$w' =$ ——	$t_i =$ ——
$c =$ ——	$c' =$ ——	$t =$ ——	$t_f =$ ——
$\therefore$ sp. h't of turpentine = ——		$\therefore$ sp. h't of turpentine = ——	
$\frac{z_1}{z_2}(178) =$ ——	$\frac{40 - t_j}{70 - t_j} =$ ——	$\frac{z_1}{z_2} =$ ——	$\frac{40 - t_j}{70 - t_j} =$ ——

2.

1st trial		2d trial	
$W_c =$ ——		$W_c =$ ——	
$W_b =$ ——	$t_i =$ ——	$W_b =$ ——	$t_i =$ ——
$\therefore W_w =$ ——	$t_f =$ ——	$W_w =$ ——	$t_f =$ ——
$W_m =$ ——	$t'_i =$ ——	$W_m =$ ——	$t'_i =$ ——
$\therefore W_t =$ ——	$t'_f =$ ——*	$W_t =$ ——	$t'_f =$ ——*

Radiation cons't (171) M = —— $n =$ —— $d =$ —— $\therefore k =$ ——

Radiation correct'n (170) $t_m =$ —— $t_j =$ —— $N =$ —— $\therefore L =$ ——

Heat capacity of cal'r and stirrer = —— of thermom'r = ——

$\therefore$ sp. h't of turpentine = —— $\therefore$ sp. h't of turpentine = ——

Problems

1. 50 gm. of ice are dropped into a brass calorimeter containing 800 gm. of water at 27°C. The calorimeter weighs 150 gm. Find the final temperature.

2. A 50 kgm. block of ice fell 30 meters. How many grams of ice were melted by the heat generated by the fall? (4.2×10^7 ergs = 1 calorie.)

3. According to very careful determinations made by Violle the quantity of heat required to raise 1 gm. of platinum from 0° to t° is given for all temperatures by the formula $Q = .0317t + .000006t^2$. Find the temperature of a Bunsen flame if a 20 gm. platinum ball dropped from the flame into 377 gm. of water at 0° raised the temperature of the water 2°.

---

* t'_f is t_f reduced to terms of thermometer No. 1. It is found by comparing No. 1 and No. 2 at about the temperature t_f. It is evident that this proceeding diminishes errors due to imperfect thermometers.

4. What should be the result of mixing 10 gm. of snow at 0° with 10 gm. of water at 35°C.?

5. The globe of a steam calorimeter is made of brass, weighs 50 gm. and has a volume of 1 liter. It contains nitrogen at a pressure of 2 atmospheres. The temperature of the chamber is 10°. What will be the increase in weight upon the admission of steam?

XXIII

EXPANSION

Theory

The true *coefficient of expansion* c of any body at any temperature is defined as *the ratio* between the volume increase, produced by an infinitely small rise in temperature, and the volume of the body at zero. Thus if V_0, V_{t_1}, and V_{t_2} represent the volumes at 0°, t_1° and t_2°, then, if t_1 and t_2 are infinitely close together,

Definition.

$$c = \frac{V_{t_2} - V_{t_1}}{(t_2 - t_1)\ V_0}. \qquad (179)$$

For hydrogen this quantity is the same for all temperatures simply by virtue of the definition of temperature; for it is the expansion of hydrogen which has been made the measure of temperature change (see p. 126). For the other gases it is constant by virtue of the law of Gay-Lussac (see p. 121). Hence, for gases which follow closely this law, equation (179) gives the correct definition of the expansion coefficient, even when t_1 and t_2 represent widely different temperatures. For such cases the volume at t°, viz. V_t, expressed in terms of the volume at 0°, viz. V_0, is [cf. (179) when $t_1 = 0$]

Coefficient is constant for gases.

$$V_t = V_0\,(1 + ct). \qquad (180)$$

The density at t°, viz. D_t, in terms of the density at 0°, viz. D_0, is found by substituting in (180) the relation $\frac{\text{Mass}}{\text{Density}} = \text{Volume}$. This gives

$$D_t = \frac{D_0}{1 + ct}. \qquad (181)$$

That c can not be constant for most liquids and solids is evident from the fact, already mentioned on p. 123, that thermometers made from liquids or solids by dividing the increase in volume between 0° and 100° into 100 equal parts, do not in general agree at intermediate temperatures with the hydrogen thermometer. For most solids, however, and for mercury, the departures are slight for temperatures below 100°. For example, a mercury in glass thermometer graduated in the manner just indicated differs from a hydrogen thermometer at no point between 0° and 100° by more than .2°. Hence, in ordinary work with these substances, at ordinary temperatures, c is usually considered constant, and equations (180) and (181) are applied precisely as in the case of gases. For high temperatures, however, this approximation can not be used.

Coefficient is nearly constant for solids and mercury.

The expansion coefficients of most liquids other than mercury increase rapidly with the temperature. For example, in passing from 0° to 40° the coefficient of turpentine increases 3.9%; of alcohol, 4.4%; of bisulphide of carbon, 4.9%; of ether, 6.6%. Between 0° and 4° water possesses the peculiar property, which is also shown by certain alloys, of contracting as the temperature rises, i.e., it has a negative coefficient. It is evident, then, that in general if t_1 and t_2 represent widely different temperatures, (179) gives the mean coefficient between t_1 and t_2, rather than the true coefficient at any particular temperature.

Coefficient not constant for most liquids.

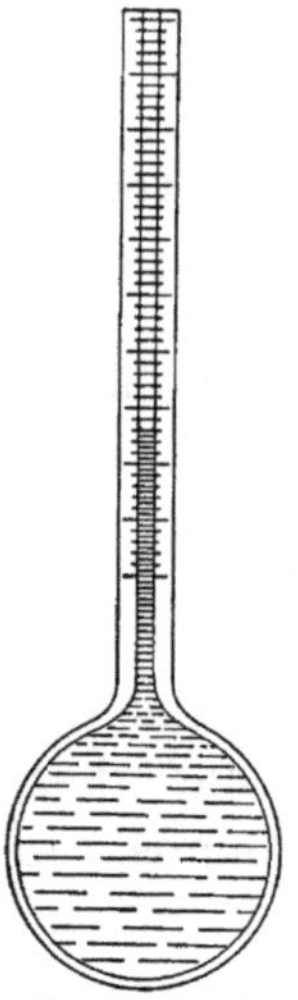
FIGURE 120

It is to be observed also that in the case of both solids and liquids (not, however, in the case of gases), the increase in volume $(V_{t_2} - V_{t_1})$ [see (179)] is so small, even when t_1 and t_2 differ from each other by as much as 100°, that the error introduced into (179) by replacing V_0 by the volume at any ordinary temperature is less than the necessary observational error in obtaining the increase $(V_{t_2} - V_{t_1})$.

V_0 replaceable by V.

The most accurate and convenient method of studying the coefficients of liquids at various temperatures is to fill a glass vessel having a large bulb and a capillary neck (see Fig. 120), to raise the temperature from t_1 to t_2 and to observe the corresponding rise l of the liquid in the neck. If a represent the area of a cross-section of the tube, then the *apparent* increase in the volume of the liquid is al. This apparent increase is, however, in reality the differences between

Method of studying expansion of liquids.

the expansions of the glass and of the liquid; for the increase in the capacity of the bulb is precisely the same as would be the increase in the volume of a solid glass vessel of the same volume as the interior of the bulb. This is evident from the consideration that a solid bulb may be conceived as made up of a series of concentric hollow bulbs, each of which expands independently of all the rest. Hence the increase in the interior capacity of each hollow shell is the same as the increase in volume of the solid bulb which fills it. Hence, if V_0 represent the interior volume of the bulb at 0° and γ, the expansion coefficient of glass, the increase in capacity of the bulb, viz. $(V_{t_2} - V_{t_1})$ is, by (179), equal to $V_0(t_2 - t_1)\gamma$. Similarly, if c be the coefficient of the liquid, the real increase in its volume is $V_0(t_2 - t_1)c$. Since, then, the apparent increase al is the difference between these quantities, there results

$$c - \gamma = \frac{al}{V_0(t_2 - t_1)}; \tag{182}$$

an equation which shows that it is impossible to obtain c from the measurement of a, l, V_0, t_1 and t_2, unless γ is already known.

A similar condition exists with respect to the volume coefficients of solids. For one of the most precise methods of measuring this quantity consists in finding the densities D_{t_1} and D_{t_2} of the solid at t_1° and t_2° by the loss of weight method (see Ex. XX), and then substituting in the equation which results from replacing the volumes in (179) by the relation $\frac{\text{Mass}}{\text{Density}}$. This gives

Measurement of the volume coefficient of solids.

$$c = \frac{(D_{t_1} - D_{t_2})D_0}{(t_2 - t_1)D_{t_1}D_{t_2}}. \tag{183}$$

But the density of a solid can not be found by the loss of weight method unless the density of the liquid in which it is immersed is known. If the expansion coefficient c of one single liquid could be accurately investigated by a method which was independent of the expansion of glass or of any other substance, this liquid could be used for determining γ for any glass bulb [see (182)], and this bulb could thereafter be used for investigating the coefficients of all other liquids. This would in turn render the method of (183) serviceable for determining the coefficients of solids.

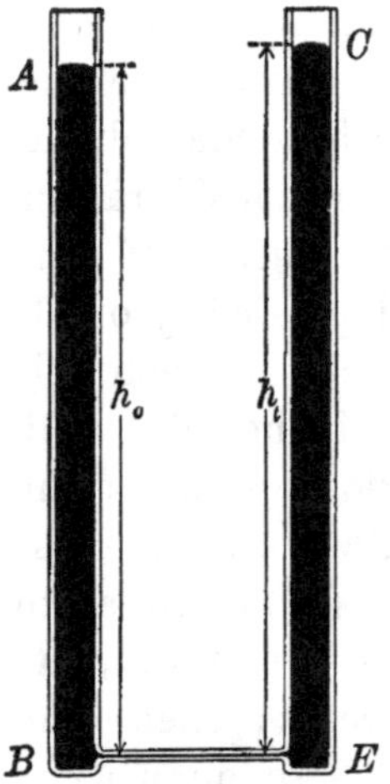

FIGURE 121

Absolute expansion of mercury.

The liquid which has been chosen for this special investigation is mercury. The method used was devised early in the last century by Dulong and Petit. Regnault afterward repeated the determinations and obtained results in close agreement with those of the earlier investigators. The principle of the method is very simple, though its application is less so. The apparatus consists essentially of two tubes *AB* and *CE* (see Fig. 121) which are connected at the bottom by a capillary tube *BE*. Tube *AB* is surrounded with melting ice and *CE* with a water-jacket of known temperature *t*. The levels at *A* and *C* must adjust themselves so that the pressures at *B* and *E* are equal; but the pressure at *B* is D_0h_0 (see Fig. 121), and that at *E* is D_th_t, D_0 and D_t being the densities of mercury at 0° and $t°$ respectively. Hence

$$h_0D_0 = h_t D_t. \qquad (184)$$

It follows from (184) and (181) that the value of *c* between 0° and $t°$ is given by

$$c = \frac{h_t - h_0}{h_0 t}. \qquad (185)$$

Regnault obtained for the mean coefficient of mercury between 0° and 100°, .0001815.

The volume (or cubical) coefficient which has been thus far discussed is the only expansion coefficient which liquids possess;

Linear expansion of solids.

but for solids there exists also a linear coefficient which is defined by

$$\lambda = \frac{l_{t_2} - l_{t_1}}{(t_2 - t_1)l_0}, \qquad (186)$$

in which l_0, l_{t_1}, and l_{t_2} are the lengths of any given line of the body at 0°, $t_1°$, and $t_2°$ respectively. It is this coefficient which is usually made the subject of measurement in the case of solids; for the cubical coefficient *c* can be deduced from it by means of the simple relation

$$c = 3\lambda. \qquad (187)$$

To prove that this relation exists, let l, l', and l'' be the three edges of a rectangular block at a temperature of $t°$. These three lengths may be expressed in terms of λ and the corresponding lengths at 0°, thus [see (186)],

Proof that $c = 3\lambda$.

$$l_t = l_0\,(1 + \lambda t),$$
$$l'_t = l'_0\,(1 + \lambda t),$$
$$l''_t = l''_0\,(1 + \lambda t).$$

Multiplication gives

$$l\,l'l'' = l_0\,l'_0\,l''_0\,(1 + 3\lambda t + 3\lambda^2 t^2 + \lambda^3 t^3); \qquad (188)$$

or

$$V_t = V_0\,(1 + 3\lambda t + 3\lambda^2 t^2 + \lambda^3 t^3). \qquad (189)$$

But since λ is always very small, scarcely ever greater than .00003, the terms in λ^2 and λ^3 are wholly negligible in comparison with the term in λ. Hence

$$V_t = V_0\,(1 + 3\lambda t). \qquad (190)$$

Comparison with (180) shows that $c = 3\lambda$. Q. E. D.

Linear expansion coefficients have for the most part been determined by one of two very simple methods. The first was introduced by Lavoisier and Laplace toward the end of the eighteenth century. It consists in placing against a rigid wall (see Fig. 122) one end of the bar which is to be heated, attaching an optical lever to the other end, and measuring the expansion precisely as the elongation of the wire was measured in Ex. VIII.

Methods of measuring linear coefficients.

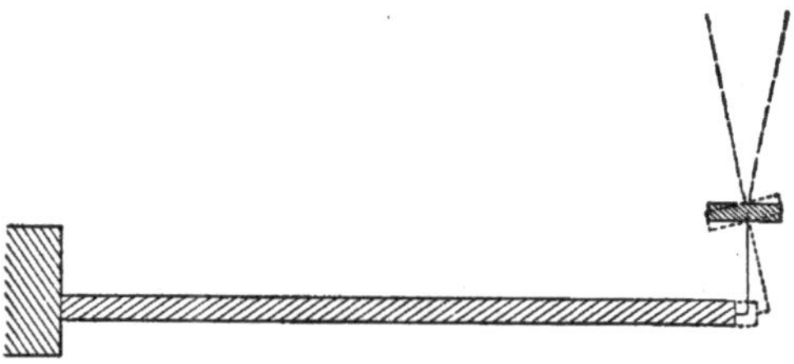

FIGURE 122

The other method, which is the one now in use at the International Bureau of Weights and Measures, consists in focusing two fixed microscopes upon fine scratches near the ends of the bar to be investigated, and measuring directly with the aid of a micrometer eyepiece and a comparison scale, the elongation produced by a given rise in temperature (see Fig. 123). Different bars of the same material show very considerable differences in expansion coefficients. Hence, in general, the numbers given in tables as the coefficients of solids must be looked upon merely as mean values.

Experiment

Object. To determine the linear coefficient of expansion of a hollow brass tube.

Place the tube *a* (see Fig. 123) in the non-conducting hair-felt covering *b*, mount it upon supports as in the figure, and focus the two micrometer microscopes *A* and *B* upon fine scratches near the two ends of *a*. In this setting see to it that the microscope is placed so that the scratch appears on that side of the field of view which is farthest from the

The preliminary adjustments.

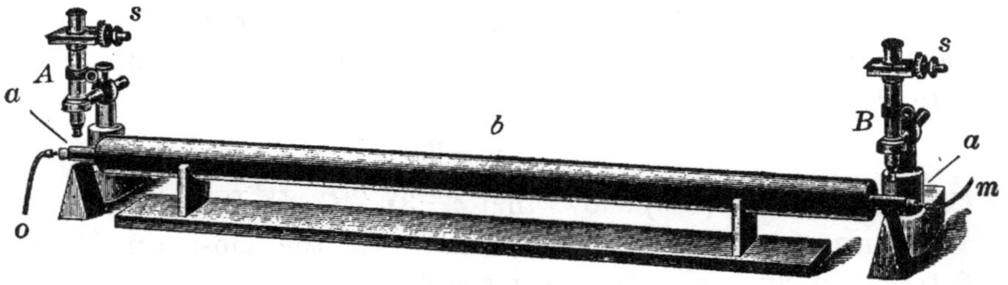

FIGURE 123

middle of the rod. This is done so that the expansion may not cause the scratches to move out of the field. It will be observed also that it takes into account the inversion of image produced by the microscope. Connect the rubber tube *m* with the water tap and take the temperature of the water by holding a thermometer in the stream which emerges from *o*. Then set the movable cross-hairs of the microscopes accurately upon the scratches and take the readings upon the scales in the micrometer eyepieces.

The microscope readings.

This is done by first rotating the micrometer screws in such a direction that the reading upon the circular head continually increases, i.e., passes from 0 toward 100, at the same time observing in what direction the movable cross-hairs pass over the field of view. If it be found that they move from right to left, then the reading of the scratch is the number of turns (i.e., teeth) and fractions of a turn (see micrometer head) through which the micrometer screw has turned in bringing the cross-hairs from the extreme *right-hand* tooth up to its present position. This is read off directly upon the toothed scale and the micrometer head. It will be observed that in order to facilitate the counting of the teeth every fifth notch is more deeply indented than the rest. If the cross-hairs had been observed to move from left to rignt, the extreme *left-hand* tooth would of course have been chosen as the point of reference. Let the recorded reading be a mean of a number of settings, and in each setting, in order to avoid the error due to the play, or back-lash, between the nut and the screw, bring the cross-hairs up to the scratch from the same side.

As soon as the two readings have been taken, turn off the current of cold water, using extreme precautions against disturbing the tube *a*, transfer the rubber connecting tube from the water tap to a steam boiler, and allow a rapid current of steam to pass through *a* until expansion ceases. Take again a set of readings at both ends. Then find the reducing factors of each microscope by focusing upon a standard scale and observing the number of turns to the millimeter. Reduce the observed expansion to millimeters. Measure the distance between the scratches by means of an ordinary meter stick. Obtain the temperature of the steam from the barometer height and the table in the Appendix. Then compute λ [see (186)].

Final adjustments.

Record

No. of screw turns to mm. in microscope $A =$ —— in $B =$ ——

1st trial	Temperature	Reading in A	Reading in B	2d trial	Temperature	Reading in A	Reading in B
Water	——	——	——	Water	——	——	——
Steam	——	——	——	Steam	——	——	——
Differences		——	——	Differences		——	——
Diff's in mm.		——	——	Diff's in mm.		——	——
$l_0 =$ ——		$l_{t_2} - l_{t_1} =$	——	$l_0 =$ ——		$l_{t_2} - l_{t_1} =$	——
∴ Expansion coef. $c =$			——	∴ Expansion coef. $c =$			——

Problems

1. In compensated pendulums the expansion of one set of rods lowers the bob, while that of another raises it. Suppose the first set to be made of iron and to have a total length of 90 cm. If the material of the second set is zinc, what must be their total length?

2. If a clock which has an uncompensated pendulum made of brass keeps correct time at 15°, how many seconds will it lose per day at 25°?

3. A surveyor's steel tape which is 10 meters long is correct at 15°. What is its error at 0°? at 100°?

4. A glass bulb of 10 cm. capacity is exactly full of mercury at 20°. How many grams of mercury will run out if it is heated to 100°?

5. A barometer provided with a brass scale correct at 15° read 735.65 mm. on a day on which the temperature was 25°. By means of the expansion coefficients of brass and mercury, find the correct barometer height at 0° upon the day in question.

6. A cubical block of steel 30 cm. on an edge floats on mercury. How far will it sink when the temperature rises from 15° to 75°?

APPENDIX

THE MICROMETER CALIPER

In the micrometer caliper (see Fig. 124) the divisions upon the scale *c* correspond to the pitch of the screw *s*. When the jaws *ab* are in contact the edge of the sleeve *d* coincides with the zero line of the scale *c*, and the zero division upon the graduated edge *d* coincides with the line *ec*. Hence, when the jaws are separated by a rotation of the milled head *h*, the whole number of screw turns of separation is read off directly upon the scale *c*, while the

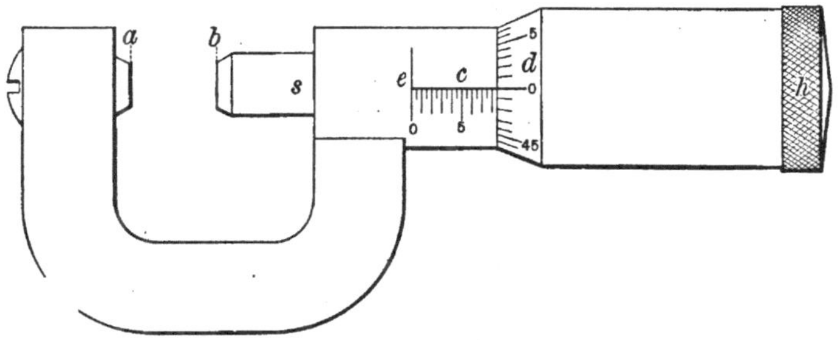

FIGURE 124

fraction of a screw turn is given by the graduation upon the edge *d*. For example, if the pitch of the screw is .5 mm., and if there are fifty divisions upon the circumference of *d*, then each of these divisions corresponds to a motion of .01 mm. at *b*. By estimating to tenths of a division upon *d*, the separation of the jaws can be read to .001 mm.

To make a measurement, place the object between the jaws *ab* and turn up the milled head *h* until, with light pressure between thumb and finger, the head slips through the fingers instead of rotating farther. In the best instruments the head is arranged to slip upon the screw as soon as a certain pressure has been reached. If this arrangement is absent, take great care not to crowd the screw. Without removing the object from *ab* read upon the scales *c* and *d* the separation of the jaws to .001 mm. Then remove the object, close the jaws, using the same pressure as before, and if the zero of *d* does not coincide exactly with the line *ec*, observe the difference in thousandths millimeters, and correct the first reading for this zero error of the caliper.

THE VERNIER CALIPER

A vernier is an auxiliary sliding scale which enables an observer to increase somewhat his accuracy in estimating the fractional portion of the last small division of the main scale. For example, it is seen at once from the main scale *DE* (Fig. 125), that the length of the rod *AB* is about 2.7 units. The use of the auxiliary scale *AC* removes all uncertainty as to this tenths place, and even makes possible an estimate in hundredths place. For *AC* is so divided that ten of its divisions are exactly equal to nine divisions on *DE*. Hence, if the zero of *AC* were back in coincidence with the mark **2** of *DE*, the mark 1 of *AC* would be just one-tenth of one of the *DE* divisions behind the mark **3** on *DE*. Similarly, 2 would be two tenths behind **4**, 3 three tenths behind **5**, and so on. Hence, if the vernier scale were moved up so as to bring its 1 mark into coincidence with the mark nearest to it on *DE*, the distance which its zero would move up beyond **2** would be one-

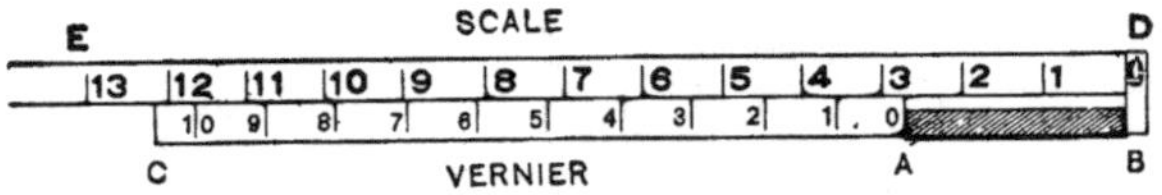

FIGURE 125

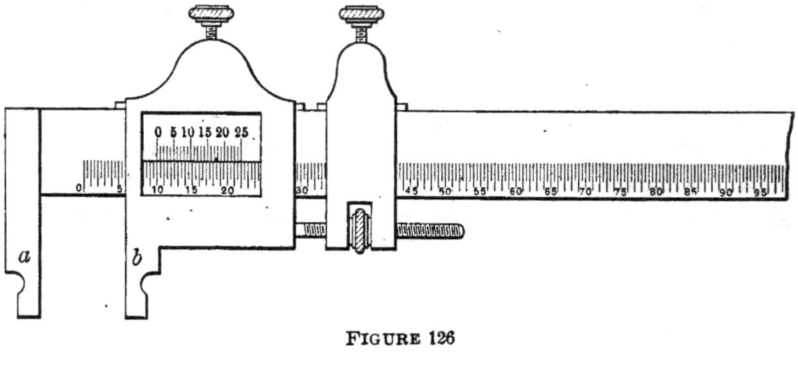

FIGURE 126

tenth of a unit. If it were moved up so as to bring the **5** into coincidence, the zero of *AC* would have moved up five tenths units beyond 2. In general, then, it is only necessary to observe which mark on the vernier AC is in coincidence with a mark on DE in order to know how many tenths the zero of AC has moved up beyond the last division passed on DE. A glance at Fig. 125 shows that 6 has already passed coincidence, while 7 has not quite reached the coincidence. The length of AB is then 2.6+. The next figure beyond the 6 is nearer to 7 than to 6, i.e., it is between 2.65 and 2.70. If then the length be recorded as 2.67, the only estimate will be in the figure 7, i.e., in hundredths instead of in tenths place as at first. Thus a vernier which has 10 divisions to 9 divisions on the main scale reads directly to tenths of the main scale divisions. In the same way a vernier which has 25 divisions equal to 24 main scale divisions reads directly to twenty-fifths of the main scale divisions. If the ratio is 50 to 49 the vernier reads to fiftieths, etc. Fig. 126 shows a vernier caliper in which the vernier has 25 divisions to 24 half-millimeter divisions on the main scale. It therefore reads to .02 mm. The body is placed between the jaws *a* and *b*, and the reading taken directly upon the scale and vernier.

TABLES

Table 1

Coefficients of Elasticity

Substances	Volume Elasticity $= \kappa$	Simple Rigidity $= n$	Young's Modulus $= Y$
Water	2.2×10^{10}	....	....
Mercury	2.6×10^{11}	....	....
Glass	$4.1 \times$ "	2.4×10^{11}	6.0×10^{11}
Brass, drawn	$10.8 \times$ "	$3.7 \times$ "	$10.8 \times$ "
Steel	$18.4 \times$ "	$8.2 \times$ "	$21.4 \times$ "
Wrought iron	$14.6 \times$ "	$7.7 \times$ "	$19.6 \times$ "
Cast iron	$9.6 \times$ "	$5.3 \times$ "	$13.5 \times$ "
Copper	$16.8 \times$ "	$4.5 \times$ "	$12.3 \times$ "
Aluminium	$5.5 \times$ "	$2.5 \times$ "	$6.5 \times$ "

Table 2

Densities

Solids

Aluminium	2.58	Iron, wrought	7.86
Brass (about)	8.5	Lead	11.3
Brick	2.1	Nickel	8.9
Copper	8.92	Oak	0.8
Cork	0.24	Pine	0.5
Diamond	3.52	Platinum	21 50
Glass (common crown)	2.6	Quartz	2.65
" (flint)	3.0-6.3	Silver	10.53
Gold	19.3	Sugar	1.6
Ice at 0° C.	0.91	Tin	7.29
Iron, cast	7.4	Zinc	7.15

Mean density of earth is 5.5270.

Liquids

Alcohol at 20° C.	0.789	Mercury	13.596
Carbon bisulphide	1.29	Sulphuric acid	1.85
Ethyl ether at 0° C.	0.735	Hydrochloric acid	1.27
Glycerine	1 26	Nitric acid	1.56
Turpentine	0.87	Olive oil	0.91

Gases at 0° C. and 76 Centimeters of Mercury Pressure

Air, dry	0.001293	Hydrogen	0.0000895
Ammonia	0.000770	Marsh gas	0.000715
Carbon dioxide	0.001974	Nitrogen	0.001257
Chlorine	0.003133	Oxygen	0.001430

Table 3

DENSITY OF DRY AIR AT TEMPERATURE t AND PRESSURE H mm. OF MERCURY

t	$H=720$	730	740	750	760	770	DIFFERENCE PER mm.
10°	.001181	.001198	.001214	.001231	.001247	.001263	16
11	1177	1194	1210	1226	1243	1259	1 2
12	1173	1189	1206	1222	1238	1255	2 3
13	1169	1185	1202	1218	1234	1250	3 5
14	1165	1181	1197	1214	1230	1246	4 6
15°	.001161	.001177	.001193	.001209	.001225	.001242	5 8
16	1157	1173	1189	1205	1221	1237	6 10
17	1153	1169	1185	1201	1217	1233	7 11
18	1149	1165	1181	1197	1213	1229	8 13
19	1145	1161	1177	1193	1209	1224	9 14
20°	.001141	.001157	.001173	.001189	.001204	.001220	15
21	1137	1153	1169	1185	1200	1216	1 2
22	1133	1149	1165	1181	1196	1212	2 3
23	1130	1145	1161	1177	1192	1208	3 4
24	1126	1141	1157	1173	1188	1204	4 6
25°	.001122	.001138	.001153	.001169	.001184	.001200	5 7
26	1118	1134	1149	1165	1180	1196	6 9
27	1114	1130	1145	1161	1176	1192	7 10
28	1110	1126	1142	1157	1172	1188	8 12
29	1107	1122	1138	1153	1169	1184	9 13
30°	.001103	.001119	.001134	.001149	.001165	.001180	

Correction for Moisture in Above Table

Dew-point	Subtract	Dew-point	Subtract	Dew-point	Subtract	Dew-point	Subtract
−10°	.000001	0°	.000003	+10°	.000006	+20°	.000010
− 8	.000002	+2	.000003	+12	.000006	+22	.000012
− 6	.000002	+4	.000004	+14	.000007	+24	.000013
− 4	.000002	+6	.000004	+16	.000008	+26	.000015
− 2	.000003	+8	.000005	+18	.000009	+28	.000016

Table 4

DENSITY OF WATER

TEMP. C°	DENSITY	TEMP. C°	DENSITY
0°	0.999884	13°	0.999443
1	0.999941	14	0.999312
2	0.999982	15	0.999173
3	1.000004	16	0.999015
3.95	1.000000	17	0.998854
4	1.000013	18	0.998667
5	1.000003	19	0.998473
6	0.999983	20	0.998272
7	0.999946	22	0.997839
8	0.999899	24	0.997380
9	0.999837	26	0.996879
10	0.999760	28	0.996344
11	0.999668	30	0.995778
12	0.999562	100	0.958860

Table 5

DENSITY OF MERCURY

TEMP. C°	DENSITY
0°	13.596
10	13.572
12	13.567
14	13.562
16	13.557
18	13.552
20	13.547
22	13.542
24	13.537
26	13.532
28	13.528
30	13.523
32	13.518
34	13.513

Table 6

Saturated Water Vapor

Showing pressure P (in mm. of mercury) and density D of aqueous vapor saturated at temperature t; or showing boiling point t of water and density D of steam corresponding to an outside pressure P.

t	P	D	t	P	D	t	P	D
—10	2.2	2.3×10^{-6}	30	31.5	30.1×10^{-6}	88.5	496.2	
— 9	2.3	2.5 "	35	41.8	39.3 "	89	505.8	
— 8	2.5	2.7 "	40	54.9	50.9 "	89.5	515.5	
— 7	2.7	2.9 "	45	71.4	65.3 "	90	525.4	428.4×10^{-6}
— 6	2.9	3.2 "	50	92.0	83.0 "	90.5	535.5	
— 5	3.2	3.4 "	55	117.5	104.6 "	91	545.7	
— 4	3.4	3.7 "	60	148.8	130.7 "	91.5	556.1	
— 3	3.7	4.0 "	65	187.0	162.1 "	92	566.7	
— 2	3.9	4.2 "	70	233.1	199.5 "	92.5	577.4	
— 1	4.2	4.5 "	71	243.6		93	588.3	
0	4.6	4.9 "	72	254.3		93.5	599.6	
1	4.9	5.2 "	73	265.4		94	610.6	
2	5.3	5.6 "	74	276.9		94.5	622.0	
3	5.7	6.0 "	75	288.8	243.7 "	95	633.6	511.1 "
4	6.1	6.4 "	75.5	294.9		95.5	645.4	
5	6.5	6.8 "	76	301.1		96	657.4	
6	7.0	7.3 "	76.5	307.4		96.5	669.5	
7	7.5	7.7 "	77	313.8		97	681.8	
8	8.0	8.2 "	77.5	320.4		97.5	694.2	
9	8.5	8.7 "	78	327.1		98	707.1	
10	9.1	9.3 "	78.5	333.8		98.2	712.3	
11	9.8	10.0 "	79	340.7		98.4	717.4	
12	10.4	10.6 "	79.5	347.7		98.6	722.6	
13	11.1	11.2 "	80	354.9	295.9 "	98.8	727.9	
14	11.9	12.0 "	80.5	362.1		99	733.2	
15	12.7	12.8 "	81	369.5		99.2	738.5	
16	13.5	13.5 "	81.5	377.0		99.4	743.8	
17	14.4	14.4 "	82	384.6		99.6	749.2	
18	15.3	15.2 "	82.5	392.4		99.8	754.7	
19	16.3	16.2 "	83	400.3		100	760.0	606.2 "
20	17.4	17.2 "	83.5	408.3		100.2	765.5	
21	18.5	18.2 "	84	416.5		100.4	771.0	
22	19.6	19.3 "	84.5	424.7		100.6	776.5	
23	20.9	20.4 "	85	433.2	357.1 "	100.8	782.1	
24	22.2	21.6 "	85.5	441.7		101	787.7	
25	23.5	22.9 "	86	450.5		102	816.0	
26	25.0	24.2 "	86.5	459.3		103	845.3	
27	26.5	25.6 "	87	468.3		105	906.4	715.4 "
28	28.1	27.0 "	87.5	477.4		107	971.1	
29	29.7	28.5 "	88	486.8		110	1075.4	840.1 "

Table 7

Reduction of Barometric Height to 0° C.

(The table corrections represent the number of millimeters to be subtracted from the observed height h. They are obtained from the formula $(.000181-.000019)ht$, the first number being the cubical expansion coefficient of mercury, the second the linear coefficient of brass.)

t	Observed Height in mm.									
	680	690	700	710	720	730	740	750	760	770
	mm.	mm.	mm.	mm.	mm.	mm.	mm.	mm.	mm.	mm.
10°	1.10	1.12	1.13	1.15	1.17	1.18	1.20	1.22	1.23	1.25
11	1.21	1.23	1.25	1.27	1.28	1.30	1.32	1.34	1.35	1.37
12	1.32	1.34	1.36	1.38	1.40	1.42	1.44	1.46	1.48	1.50
13	1.43	1.45	1.47	1.50	1.52	1.54	1.56	1.58	1.60	1.62
14	1.54	1.56	1.59	1.61	1.63	1.66	1.68	1.70	1.72	1.75
15	1.65	1.68	1.70	1.73	1.75	1.77	1.80	1.82	1.85	1.87
16	1.76	1.79	1.81	1.84	1.87	1.89	1.92	1.94	1.97	2.00
17	1.87	1.90	1.93	1.96	1.98	2.01	2.04	2.07	2.09	2.12
18	1.98	2.01	2.04	2.07	2.10	2.13	2.16	2.19	2.22	2.25
19	2.09	2.12	2.15	2.19	2.22	2.25	2.28	2.31	2.34	2.37
20	2.20	2.24	2.27	2.30	2.33	2.37	2.40	2.43	2.46	2.49
21	2.31	2.35	2.38	2.42	2.45	2.48	2.52	2.55	2.59	2.62
22	2.42	2.46	2.49	2.53	2.57	2.60	2.64	2.67	2.71	2.74
23	2.53	2.57	2.61	2.65	2.68	2.72	2.76	2.79	2.83	2.87
24	2.64	2.68	2.72	2.76	2.80	2.84	2.88	2.92	2.95	2.99
25	2.75	2.79	2.84	2.88	2.92	2.96	3.00	3.04	3.08	3.12

Table 8

Capillary Depression of Mercury in Glass

Diameter	Height of the Meniscus in mm.							
	0.4	0.6	0.8	1.0	1.2	1.4	1.6	1.8
mm.	mm.	mm.	mm.	mm.	mm.	mm.	mm.	mm.
4	0.83	1.22	1.54	1.98	2.37			
5	0.47	0.65	0.86	1.19	1.45	1.80		
6	0.27	0.41	0.56	0.78	0.98	1.21	1.43	
7	0.18	0.28	0.40	0.53	0.67	0.82	0.97	1.13
8		0.20	0.29	0.38	0.46	0.56	0.65	0.77
9		0.15	0.21	0.28	0.33	0.40	0.46	0.52
10			0.15	0.20	0.25	0.29	0.33	0.37
11			0.10	0.14	0.18	0.21	0.24	0.27
12			0.07	0.10	0.13	0.15	0.18	0.19
13			0.04	0.07	0.10	0.12	0.13	0.14

Table 9

SURFACE TENSIONS

in dynes per cm.

Alcohol	at 20°	25.5	Olive Oil	at 20°	31.7
Benzine	at 15°	28.8	Petroleum	at 20°	23.9
Glycerine	at 17°	63.1	Water	at 0°	76.6
Mercury	at 20°	450.0	Water	at 20°	74.0

Table 10

AVERAGE SPECIFIC HEATS

Alcohol	at 40°	0.648	Lead	0°-100°	0.0315
Aluminium	0°-100°	0.2185	Mercury	20°- 50°	0.0333
Brass		0.094	Paraffine		0.683
Copper	0°-100°	0.095	Platinum	0°-100°	0.0323
German-silver		0.0946	Silver	0°-100°	0.0568
Glass		0.20	Steel		0.118
Gold		0.0316	Tin	0°-100°	0.0559
Ice		0.504	Turpentine		0.467
Iron	0°-100°	0.1130	Zinc		0.0935

Table 11

AVERAGE COEFFICIENTS OF EXPANSION BETWEEN 0° AND 100° C.

Cubical

Glass	0.000025	Mercury	0.0001815	Turpentine	0.00105

Linear

Aluminium	0.000023	Iron (soft)	0.000012	Silver	0.000019
Brass	0.000018	Iron (cast)	0.0000105	Steel	0.000011
Copper	0.000017	Lead	0.000029	Tin	0.000022
Gold	0.000014	Platinum	0.000009	Zinc	0.000029

Table 12

IMPORTANT NUMBERS

$\pi = 3.1416$. $\pi^2 = 9.8696$. $\frac{1}{\pi} = 0.31831$. Logarithm $\pi = .49715$.

Base of the natural system of logarithms $\epsilon = 2.7183$.

1 inch = 25.4 millimeters. 1 meter = 39.37 inches. 1 mile = 1.609 kilometers.

1 kilogram = 2.2 pounds. 1 ounce = 28.35 grams. 1 grain = 64.8 milligrams.

Mechanical equivalent of 1 calorie (15°) $= 4.19 \times 10^7$ ergs.

NATURAL SINES

Angle	.0	.1	.2	.3	.4	.5	.6	.7	.8	.9	Complement Difference		
0°	0.0000	0017	0035	0052	0070	0087	0105	0122	0140	0157	0175	**89°**	
1	0175	0192	0209	0227	0244	0262	0279	0297	0314	0332	0349	**88**	
2	0349	0366	0384	0401	0419	0436	0454	0471	0488	0506	0523	**87**	
3	0523	0541	0558	0576	0593	0610	0628	0645	0663	0680	0698	**86**	
4	0698	0715	0732	0750	0767	0785	0802	0819	0837	0854	0872	**85**	
5	0.0872	0889	0906	0924	0941	0958	0976	0993	1011	1028	1045	**84**	
6	1045	1063	1080	1097	1115	1132	1149	1167	1184	1201	1219	**83**	
7	1219	1236	1253	1271	1288	1305	1323	1340	1357	1374	1392	**82**	
8	1392	1409	1426	1444	1461	1478	1495	1513	1530	1547	1564	**81**	
9	1564	1582	1599	1616	1633	1650	1668	1685	1702	1719	1736	**80**	
10	0.1736	1754	1771	1788	1805	1822	1840	1857	1874	1891	1908	**79**	
11	1908	1925	1942	1959	1977	1994	2011	2028	2045	2062	2079	**78**	
12	2079	2096	2113	2130	2147	2164	2181	2198	2215	2233	2250	**77**	17
13	2250	2267	2284	2300	2317	2334	2351	2368	2385	2402	2419	**76**	
14	2419	2436	2453	2470	2487	2504	2521	2538	2554	2571	2588	**75**	
15	0.2588	2605	2622	2639	2656	2672	2689	2706	2723	2740	2756	**74**	
16	2756	2773	2790	2807	2823	2840	2857	2874	2890	2907	2924	**73**	
17	2924	2940	2957	2974	2990	3007	3024	3040	3057	3074	3090	**72**	
18	3090	3107	3123	3140	3156	3173	3190	3206	3223	3239	3256	**71**	
19	3256	3272	3289	3305	3322	3338	3355	3371	3387	3404	3420	**70**	
20	0.3420	3437	3453	3469	3486	3502	3518	3535	3551	3567	3584	**69**	
21	3584	3600	3616	3633	3649	3665	3681	3697	3714	3730	3746	**68**	
22	3746	3762	3778	3795	3811	3827	3843	3859	3875	3891	3907	**67**	
23	3907	3923	3939	3955	3971	3987	4003	4019	4035	4051	4067	**66**	16
24	4067	4083	4099	4115	4131	4147	4163	4179	4195	4210	4226	**65**	
25	0.4226	4242	4258	4274	4289	4305	4321	4337	4352	4368	4384	**64**	
26	4384	4399	4415	4431	4446	4462	4478	4493	4509	4524	4540	**63**	
27	4540	4555	4571	4586	4602	4617	4633	4648	4664	4679	4695	**62**	
28	4695	4710	4726	4741	4756	4772	4787	4802	4818	4833	4848	**61**	
29	4848	4863	4879	4894	4909	4924	4939	4955	4970	4985	5000	**60**	
30	0.5000	5015	5030	5045	5060	5075	5090	5105	5120	5135	5150	**59**	15
31	5150	5165	5180	5195	5210	5225	5240	5255	5270	5284	5299	**58**	
32	5299	5314	5329	5344	5358	5373	5388	5402	5417	5432	5446	**57**	
33	5446	5461	5476	5490	5505	5519	5534	5548	5563	5577	5592	**56**	
34	5592	5606	5621	5635	5650	5664	5678	5693	5707	5721	5736	**55**	
35	0.5736	5750	5764	5779	5793	5807	5821	5835	5850	5864	5878	**54**	
36	5878	5892	5906	5920	5934	5948	5962	5976	5990	6004	6018	**53**	14
37	6018	6032	6046	6060	6074	6088	6101	6115	6129	6143	6157	**52**	
38	6157	6170	6184	6198	6211	6225	6239	6252	6266	6280	6293	**51**	
39	6293	6307	6320	6334	6347	6361	6374	6388	6401	6414	6428	**50**	
40	0.6428	6441	6455	6468	6481	6494	6508	6521	6534	6547	6561	**49**	
41	6561	6574	6587	6600	6613	6626	6639	6652	6665	6678	6691	**48**	13
42	6691	6704	6717	6730	6743	6756	6769	6782	6794	6807	6820	**47**	
43	6820	6833	6845	6858	6871	6884	6896	6909	6921	6934	6947	**46**	
44°	6947	6959	6972	6984	6997	7009	7022	7034	7046	7059	7071	**45°**	
Complement		.9	.8	.7	.6	.5	.4	.3	.2	.1	.0	Angle	

NATURAL COSINES

NATURAL SINES

Angle	.0	.1	.2	.3	.4	.5	.6	.7	.8	.9	Complement Difference		
45°	0.7071	7083	7096	7108	7120	7133	7145	7157	7169	7181	7193	**44°**	
46	7193	7206	7218	7230	7242	7254	7266	7278	7290	7302	7314	**43**	12
47	7314	7325	7337	7349	7361	7373	7385	7396	7408	7420	7431	**42**	
48	7431	7443	7455	7466	7478	7490	7501	7513	7524	7536	7547	**41**	
49	7547	7559	7570	7581	7593	7604	7615	7627	7638	7649	7660	**40**	
50	0.7660	7672	7683	7694	7705	7716	7727	7738	7749	7760	7771	**39**	
51	7771	7782	7793	7804	7815	7826	7837	7848	7859	7869	7880	**38**	11
52	7880	7891	7902	7912	7923	7934	7944	7955	7965	7976	7986	**37**	
53	7986	7997	8007	8018	8028	8039	8049	8059	8070	8080	8090	**36**	
54	8090	8100	8111	8121	8131	8141	8151	8161	8171	8181	8192	**35**	
55	0.8192	8202	8211	8221	8231	8241	8251	8261	8271	8281	8290	**34**	10
56	8290	8300	8310	8320	8329	8339	8348	8358	8368	8377	8387	**33**	
57	8387	8396	8406	8415	8425	8434	8443	8453	8462	8471	8480	**32**	
58	8480	8490	8499	8508	8517	8526	8536	8545	8554	8563	8572	**31**	
59	8572	8581	8590	8599	8607	8616	8625	8634	8643	8652	8660	**30**	9
60	0.8660	8669	8678	8686	8695	8704	8712	8721	8729	8738	8746	**29**	
61	8746	8755	8763	8771	8780	8788	8796	8805	8813	8821	8829	**28**	
62	8829	8838	8846	8854	8862	8870	8878	8886	8894	8902	8910	**27**	8
63	8910	8918	8926	8934	8942	8949	8957	8965	8973	8980	8988	**26**	
64	8988	8996	9003	9011	9018	9026	9033	9041	9048	9056	9063	**25**	
65	0.9063	9070	9078	9085	9092	9100	9107	9114	9121	9128	9135	**24**	
66	9135	9143	9150	9157	9164	9171	9178	9184	9191	9198	9205	**23**	7
67	9205	9212	9219	9225	9232	9239	9245	9252	9259	9265	9272	**22**	
68	9272	9278	9285	9291	9298	9304	9311	9317	9323	9330	9336	**21**	
69	9336	9342	9348	9354	9361	9367	9373	9379	9385	9391	9397	**20**	6
70	0.9397	9403	9409	9415	9421	9426	9432	9438	9444	9449	9455	**19**	
71	9455	9461	9466	9472	9478	9483	9489	9494	9500	9505	9511	**18**	
72	9511	9516	9521	9527	9532	9537	9542	9548	9553	9558	9563	**17**	
73	9563	9568	9573	9578	9583	9588	9593	9598	9603	9608	9613	**16**	5
74	9613	9617	9622	9627	9632	9636	9641	9646	9650	9655	9659	**15**	
75	0.9659	9664	9668	9673	9677	9681	9686	9690	9694	9699	9703	**14**	
76	9703	9707	9711	9715	9720	9724	9728	9732	9736	9740	9744	**13**	4
77	9744	9748	9751	9755	9759	9763	9767	9770	9774	9778	9781	**12**	
78	9781	9785	9789	9792	9796	9799	9803	9806	9810	9813	9816	**11**	
79	9816	9820	9823	9826	9829	9833	9836	9839	9842	9845	9848	**10**	
80	0.9848	9851	9854	9857	9860	9863	9866	9869	9871	9874	9877	**9**	3
81	9877	9880	9882	9885	9888	9890	9893	9895	9898	9900	9903	**8**	
82	9903	9905	9907	9910	9912	9914	9917	9919	9921	9923	9925	**7**	
83	9925	9928	9930	9932	9934	9936	9938	9940	9942	9943	9945	**6**	2
84	9945	9947	9949	9951	9952	9954	9956	9957	9959	9960	9962	**5**	
85	0.9962	9963	9965	9966	9968	9969	9971	9972	9973	9974	9976	**4**	
86	9976	9977	9978	9979	9980	9981	9982	9983	9984	9985	9986	**3**	1
87	9986	9987	9988	9989	9990	9990	9991	9992	9993	9993	9994	**2**	
88	9994	9995	9995	9996	9996	9997	9997	9997	9998	9998	9998	**1**	
89°	9998	9999	9999	9999	9999	1.0000	1.0000	1.0000	1.0000	1.0000	1.0000	**0°**	0
Complement		.9	.8	.7	.6	.5	.4	.3	.2	.1	.0	Angle	

NATURAL COSINES

NATURAL TANGENTS

Angle	.0	.1	.2	.3	.4	.5	.6	.7	.8	.9	Complement		Difference
0°	0.0000	0017	0035	0052	0070	0087	0105	0122	0140	0157	0175	**89°**	
1	0175	0192	0209	0227	0244	0262	0279	0297	0314	0332	0349	**88**	
2	0349	0367	0384	0402	0419	0437	0454	0472	0489	0507	0524	**87**	
3	0524	0542	0559	0577	0594	0612	0629	0647	0664	0682	0699	**86**	
4	0699	0717	0734	0752	0769	0787	0805	0822	0840	0857	0875	**85**	
5	0 0875	0892	0910	0928	0945	0963	0981	0998	1016	1033	1051	**84**	
6	1051	1069	1086	1104	1122	1139	1157	1175	1192	1210	1228	**83**	
7	1228	1246	1263	1281	1299	1317	1334	1352	1370	1388	1405	**82**	
8	1405	1423	1441	1459	1477	1495	1512	1530	1548	1566	1584	**81**	
9	1584	1602	1620	1638	1655	1673	1691	1709	1727	1745	1763	**80**	
10	0.1763	1781	1799	1817	1835	1853	1871	1890	1908	1926	1944	**79**	18
11	1944	1962	1980	1998	2016	2035	2053	2071	2089	2107	2126	**78**	
12	2126	2144	2162	2180	2199	2217	2235	2254	2272	2290	2309	**77**	
13	2309	2327	2345	2364	2382	2401	2419	2438	2456	2475	2493	**76**	
14	2493	2512	2530	2549	2568	2586	2605	2623	2642	2661	2679	**75**	
15	0.2679	2698	2717	2736	2754	2774	2792	2811	2830	2849	2867	**74**	
16	2867	2886	2905	2924	2943	2962	2981	3000	3019	3038	3057	**73**	19
17	3057	3076	3096	3115	3134	3153	3172	3191	3211	3230	3249	**72**	
18	3249	3269	3288	3307	3327	3346	3365	3385	3404	3424	3443	**71**	
19	3443	3463	3482	3502	3522	3541	3561	3581	3600	3620	3640	**70**	
20	0.3640	3659	3679	3699	3719	3739	3759	3779	3799	3819	3839	**69**	
21	3839	3859	3879	3899	3919	3939	3959	3979	4000	4020	4040	**68**	20
22	4040	4061	4081	4101	4122	4142	4163	4183	4204	4224	4245	**67**	
23	4245	4265	4286	4307	4327	4348	4369	4390	4411	4431	4452	**66**	
24	4452	4473	4494	4515	4536	4557	4578	4599	4621	4642	4663	**65**	21
25	0.4663	4684	4706	4727	4748	4770	4791	4813	4834	4856	4877	**64**	
26	4877	4899	4921	4942	4964	4986	5008	5029	5051	5073	5095	**63**	
27	5095	5117	5139	5161	5184	5206	5228	5250	5272	5295	5317	**62**	22
28	5317	5340	5362	5384	5407	5430	5452	5475	5498	5520	5543	**61**	
29	5543	5566	5589	5612	5635	5658	5681	5704	5727	5750	5774	**60**	23
30	0.5774	5797	5820	5844	5867	5890	5914	5938	5961	5985	6099	**59**	
31	6009	6032	6056	6080	6104	6128	6152	6176	6200	6224	6249	**58**	24
32	6249	6273	6297	6322	6346	6371	6395	6420	6445	6469	6494	**57**	
33	6494	6519	6544	6569	6594	6619	6644	6669	6694	6720	6745	**56**	25
34	6745	6771	6796	6822	6847	6873	6899	6924	6950	6976	7002	**55**	
35	0.7002	7028	7054	7080	7107	7133	7159	7186	7212	7239	7265	**54**	26
36	7265	7292	7319	7346	7373	7400	7427	7454	7481	7508	7536	**53**	27
37	7536	7563	7590	7618	7646	7673	7701	7729	7757	7785	7813	**52**	28
38	7813	7841	7869	7898	7926	7954	7983	8012	8040	8069	8098	**51**	28
39	8098	8127	8156	8185	8214	8243	8273	8302	8332	8361	8391	**50**	29
40	0.8391	8421	8451	8481	8511	8541	8571	8601	8632	8662	8693	**49**	30
41	8693	8724	8754	8785	8816	8847	8878	8910	8941	8972	9004	**48**	31
42	9004	9036	9067	9099	9131	9163	9195	9228	9260	9293	9325	**47**	32
43	9325	9358	9391	9424	9557	9490	9523	9556	9590	9623	9657	**46**	33
44°	9657	9691	9725	9759	9793	9827	9861	9896	9930	9965	1.0000	**45°**	34
Complement		.9	.8	.7	.6	.5	.4	.3	.2	.1	.0	Angle	

NATURAL COTANGENTS

Natural Tangents

Angle	.0	.1	.2	.3	.4	.5	.6	.7	.8	.9	Dif.
45°	1.0000	1.0035	1.0070	1.0105	1.0141	1.0176	1.0212	1.0247	1.0283	1.0319	36
46	1.0355	1.0392	1.0428	1.0464	1.0501	1.0538	1.0575	1.0612	1.0649	1.0686	37
47	1.0724	1.0761	1.0799	1.0837	1.0875	1.0913	1.0951	1.0990	1.1028	1.1067	38
48	1.1106	1.1145	1.1184	1.1224	1.1263	1.1303	1.1343	1.1383	1.1423	1.1463	40
49	1.1504	1.1544	1.1585	1.1626	1.1667	1.1708	1.1750	1.1792	1.1833	1.1875	41
50	1.1918	1.1960	1.2002	1.2045	1.2088	1.2131	1.2174	1.2218	1.2261	1.2305	43
51	1.2349	1.2393	1.2437	1.2482	1.2527	1.2572	1.2617	1.2662	1.2708	1.2753	45
52	1.2799	1.2846	1.2892	1.2938	1.2985	1.3032	1.3079	1.3127	1.3175	1.3222	47
53	1.3270	1.3319	1.3367	1.3416	1.3465	1.3514	1.3564	1.3613	1.3663	1.3713	49
54	1.3764	1.3814	1.3865	1.3916	1.3968	1.4019	1.4071	1.4124	1.4176	1.4229	52
55	1.4281	1.4335	1.4388	1.4442	1.4496	1.4550	1.4605	1.4659	1.4715	1.4770	54
56	1.4826	1.4882	1.4938	1.4994	1.5051	1.5108	1.5166	1.5224	1.5282	1.5340	57
57	1.5399	1.5458	1.5517	1.5577	1.5637	1.5697	1.5757	1.5818	1.5880	1.5941	60
58	1.6003	1.6066	1.6128	1.6191	1.6255	1.6319	1.6383	1.6447	1.6512	1.6577	64
59	1.6643	1.6709	1.6775	1.6842	1.6909	1.6977	1.7045	1.7113	1.7182	1.7251	68
60	1.7321	1.7391	1.7461	1.7532	1.7603	1.7675	1.7747	1.7820	1.7893	1.7966	72
61	1.8040	1.8115	1.8190	1.8265	1.8341	1.8418	1.8495	1.8572	1.8650	1.8728	77
62	1.8807	1.8887	1.8967	1.9047	1.9128	1.9210	1.9292	1.9375	1.9458	1.9542	82
63	1.9626	1.9711	1.9797	1.9883	1.9970	2.0057	2.0145	2.0233	2.0323	2.0413	88
64	2.0503	2.0594	2.0686	2.0778	2.0872	2.0965	2.1060	2.1155	2.1251	2.1348	94
65	2.145	2.154	2.164	2.174	2.184	2.194	2.204	2.215	2.225	2.236	10
66	2.246	2.257	2.267	2.278	2.289	2.300	2.311	2.322	2.333	2.344	11
67	2.356	2.367	2.379	2.391	2.402	2.414	2.426	2.438	2.450	2.463	12
68	2.475	2.488	2.500	2.513	2.526	2.539	2.552	2.565	2.578	2.592	13
69	2.605	2.619	2.633	2.646	2.660	2.675	2.689	2.703	2.718	2.733	14
70	2.747	2.762	2.778	2.793	2.808	2.824	2.840	2.856	2.872	2.888	16
71	2.904	2.921	2.937	2.954	2.971	2.989	3.006	3.024	3.042	3.060	17
72	3.078	3.096	3.115	3.133	3.152	3.172	3.191	3.211	3.230	3.250	19
73	3.271	3.291	3.312	3.333	3.354	3.376	3.398	3.420	3.442	3.465	22
74	3.487	3.511	3.534	3.558	3.582	3.606	3.630	3.655	3.681	3.700	25
75	3.732	3.758	3.785	3.812	3.839	3.867	3.895	3.923	3.952	3.981	28
76	4.011	4.041	4.071	4.102	4.134	4.165	4.198	4.230	4.264	4.297	32
77	4.331	4.366	4.402	4.437	4.474	4.511	4.548	4.586	4.625	4.665	37
78	4.705	4.745	4.787	4.829	4.872	4.915	4.959	5.005	5.050	5.097	44
79	5.145	5.193	5.242	5.292	5.343	5.396	5.449	5.503	5.558	5.614	52
80	5.67	5.73	5.79	5.85	5.91	5.98	6.04	6.11	6.17	6.24	7
81	6.31	6.39	6.46	6.54	6.61	6.69	6.77	6.85	6.94	7.03	8
82	7.12	7.21	7.30	7.40	7.49	7.60	7.70	7.81	7.92	8.03	10
83	8.14	8.26	8.39	8.51	8.64	8.78	8.92	9.06	9.21	9.36	14
84	9.51	9.68	9.84	10.0	10.2	10.4	10.6	10.8	11.0	11.2	
85	11.4	11.7	11.9	12.2	12.4	12.7	13.0	13.3	13.6	14.0	3
86	14.3	14.7	15.1	15.5	15.9	16.3	16.8	17.3	17.9	18.5	6
87	19.1	19.7	20.4	21.2	22.0	22.9	23.9	24.9	26.0	27.3	
88	28.6	30.1	31.8	33.7	35.8	38.2	40.9	44.1	47.7	52.1	
89°	57.	64.	72.	82.	95.	115.	143.	191.	286.	573.	
Angle	.0	.1	.2	.3	.4	.5	.6	.7	.8	.9	

Natural Tangents

LOGARITHMS

	0	1	2	3	4	5	6	7	8	9	1	2	3	4	5	6	7	8	9
10	0000	0043	0086	0128	0170	0212	0253	0294	0334	0374	4	8	12	17	21	25	29	33	37
11	0414	0453	0492	0531	0569	0607	0645	0682	0719	0755	4	8	11	15	19	23	26	30	34
12	0792	0828	0864	0899	0934	0969	1004	1038	1072	1106	3	7	10	14	17	21	24	28	31
13	1139	1173	1206	1239	1271	1303	1335	1367	1399	1430	3	6	10	13	16	19	23	26	29
14	1461	1492	1523	1553	1584	1614	1644	1673	1703	1732	3	6	9	12	15	18	21	24	27
15	1761	1790	1818	1847	1875	1903	1931	1959	1987	2014	3	6	8	11	14	17	20	22	25
16	2041	2068	2095	2122	2148	2175	2201	2227	2253	2279	3	5	8	11	13	16	18	21	24
17	2304	2330	2355	2380	2405	2430	2455	2480	2504	2529	2	5	7	10	12	15	17	20	22
18	2553	2577	2601	2625	2648	2672	2695	2718	2742	2765	2	5	7	9	12	14	16	19	21
19	2788	2810	2833	2856	2878	2900	2923	2945	2967	2989	2	4	7	9	11	13	16	18	20
20	3010	3032	3054	3075	3096	3118	3139	3160	3181	3201	2	4	6	8	11	13	15	17	19
21	3222	3243	3263	3284	3304	3324	3345	3365	3385	3404	2	4	6	8	10	12	14	16	18
22	3424	3444	3464	3483	3502	3522	3541	3560	3579	3598	2	4	6	8	10	12	14	15	17
23	3617	3636	3655	3674	3692	3711	3729	3747	3766	3784	2	4	6	7	9	11	13	15	17
24	3802	3820	3838	3856	3874	3892	3909	3927	3945	3962	2	4	5	7	9	11	12	14	16
25	3979	3997	4014	4031	4048	4065	4082	4099	4116	4133	2	3	5	7	9	10	12	14	15
26	4150	4166	4183	4200	4216	4232	4249	4265	4281	4298	2	3	5	7	8	10	11	13	15
27	4314	4330	4346	4362	4378	4393	4409	4425	4440	4456	2	3	5	6	8	9	11	13	14
28	4472	4487	4502	4518	4533	4548	4564	4579	4594	4609	2	3	5	6	8	9	11	12	14
29	4624	4639	4654	4669	4683	4698	4713	4728	4742	4757	1	3	4	6	7	9	10	12	13
30	4771	4786	4800	4814	4829	4843	4857	4871	4886	4900	1	3	4	6	7	9	10	11	13
31	4914	4928	4942	4955	4969	4983	4997	5011	5024	5038	1	3	4	6	7	8	10	11	12
32	5051	5065	5079	5092	5105	5119	5132	5145	5159	5172	1	3	4	5	7	8	9	11	12
33	5185	5198	5211	5224	5237	5250	5263	5276	5289	5302	1	3	4	5	6	8	9	10	12
34	5315	5328	5340	5353	5366	5378	5391	5403	5416	5428	1	3	4	5	6	8	9	10	11
35	5441	5453	5465	5478	5490	5502	5514	5527	5539	5551	1	2	4	5	6	7	9	10	11
36	5563	5575	5587	5599	5611	5623	5635	5647	5658	5670	1	2	4	5	6	7	8	10	11
37	5682	5694	5705	5717	5729	5740	5752	5763	5775	5786	1	2	3	5	6	7	8	9	10
38	5798	5809	5821	5832	5843	5855	5866	5877	5888	5899	1	2	3	5	6	7	8	9	10
39	5911	5922	5933	5944	5955	5966	5977	5988	5999	6010	1	2	3	4	5	7	8	9	10
40	6021	6031	6042	6053	6064	6075	6085	6096	6107	6117	1	2	3	4	5	6	8	9	10
41	6128	6138	6149	6160	6170	6180	6191	6201	6212	6222	1	2	3	4	5	6	7	8	9
42	6232	6243	6253	6263	6274	6284	6294	6304	6314	6325	1	2	3	4	5	6	7	8	9
43	6335	6345	6355	6365	6375	6385	6395	6405	6415	6425	1	2	3	4	5	6	7	8	9
44	6435	6444	6454	6464	6474	6484	6493	6503	6513	6522	1	2	3	4	5	6	7	8	9
45	6532	6542	6551	6561	6571	6580	6590	6599	6609	6618	1	2	3	4	5	6	7	8	9
46	6628	6637	6646	6656	6665	6675	6684	6693	6702	6712	1	2	3	4	5	6	7	7	8
47	6721	6730	6739	6749	6758	6767	6776	6785	6794	6803	1	2	3	4	5	5	6	7	8
48	6812	6821	6830	6839	6848	6857	6866	6875	6884	6893	1	2	3	4	4	5	6	7	8
49	6902	6911	6920	6928	6937	6946	6955	6964	6972	6981	1	2	3	4	4	5	6	7	8
50	6990	6998	7007	7016	7024	7033	7042	7050	7059	7067	1	2	3	3	4	5	6	7	8
51	7076	7084	7093	7101	7110	7118	7126	7135	7143	7152	1	2	3	3	4	5	6	7	8
52	7160	7168	7177	7185	7193	7202	7210	7218	7226	7235	1	2	2	3	4	5	6	7	7
53	7243	7251	7259	7267	7275	7284	7292	7300	7308	7316	1	2	2	3	4	5	6	6	7
54	7324	7332	7340	7348	7356	7364	7372	7380	7388	7396	1	2	2	3	4	5	6	6	7

LOGARITHMS

	0	1	2	3	4	5	6	7	8	9	1	2	3	4	5	6	7	8	9
55	7404	7412	7419	7427	7435	7443	7451	7459	7466	7474	1	2	2	3	4	5	5	6	7
56	7482	7490	7497	7505	7513	7520	7528	7536	7543	7551	1	2	2	3	4	5	5	6	7
57	7559	7566	7574	7582	7589	7597	7604	7612	7619	7627	1	2	2	3	4	5	5	6	7
58	7634	7642	7649	7657	7664	7672	7679	7686	7694	7701	1	1	2	3	4	4	5	6	7
59	7709	7716	7723	7731	7738	7745	7752	7760	7767	7774	1	1	2	3	4	4	5	6	7
60	7782	7789	7796	7803	7810	7818	7825	7832	7839	7846	1	1	2	3	4	4	5	6	6
61	7853	7860	7868	7875	7882	7889	7896	7903	7910	7917	1	1	2	3	4	4	5	6	6
62	7924	7931	7938	7945	7952	7959	7966	7973	7980	7987	1	1	2	3	3	4	5	6	6
63	7993	8000	8007	8014	8021	8028	8035	8041	8048	8055	1	1	2	3	3	4	5	5	6
64	8062	8069	8075	8082	8089	8096	8102	8109	8116	8122	1	1	2	3	3	4	5	5	6
65	8129	8136	8142	8149	8156	8162	8169	8176	8182	8189	1	1	2	3	3	4	5	5	6
66	8195	8202	8209	8215	8222	8228	8235	8241	8248	8254	1	1	2	3	3	4	5	5	6
67	8261	8267	8274	8280	8287	8293	8299	8306	8312	8319	1	1	2	3	3	4	5	5	6
68	8325	8331	8338	8344	8351	8357	8363	8370	8376	8382	1	1	2	3	3	4	4	5	6
69	8388	8395	8401	8407	8414	8420	8426	8432	8439	8445	1	1	2	2	3	4	4	5	6
70	8451	8457	8463	8470	8476	8482	8488	8494	8500	8506	1	1	2	2	3	4	4	5	6
71	8513	8519	8525	8531	8537	8543	8549	8555	8561	8567	1	1	2	2	3	4	4	5	5
72	8573	8579	8585	8591	8597	8603	8609	8615	8621	8627	1	1	2	2	3	4	4	5	5
73	8633	8639	8645	8651	8657	8663	8669	8675	8681	8686	1	1	2	2	3	4	4	5	5
74	8692	8698	8704	8710	8716	8722	8727	8733	8739	8745	1	1	2	2	3	4	4	5	5
75	8751	8756	8762	8768	8774	8779	8785	8791	8797	8802	1	1	2	2	3	3	4	5	5
76	8808	8814	8820	8825	8831	8837	8842	8848	8854	8859	1	1	2	2	3	3	4	5	5
77	8865	8871	8876	8882	8887	8893	8899	8904	8910	8915	1	1	2	2	3	3	4	4	5
78	8921	8927	8932	8938	8943	8949	8954	8960	8965	8971	1	1	2	2	3	3	4	4	5
79	8976	8982	8987	8993	8998	9004	9009	9015	9020	9025	1	1	2	2	3	3	4	4	5
80	9031	9036	9042	9047	9053	9058	9063	9069	9074	9079	1	1	2	2	3	3	4	4	5
81	9085	9090	9096	9101	9106	9112	9117	9122	9128	9133	1	1	2	2	3	3	4	4	5
82	9138	9143	9149	9154	9159	9165	9170	9175	9180	9186	1	1	2	2	3	3	4	4	5
83	9191	9196	9201	9206	9212	9217	9222	9227	9232	9238	1	1	2	2	3	3	4	4	5
84	9243	9248	9253	9258	9263	9269	9274	9279	9284	9289	1	1	2	2	3	3	4	4	5
85	9294	9299	9304	9309	9315	9320	9325	9330	9335	9340	1	1	2	2	3	3	4	4	5
86	9345	9350	9355	9360	9365	9370	9375	9380	9385	9390	1	1	2	2	3	3	4	4	5
87	9395	9400	9405	9410	9415	9420	9425	9430	9435	9440	0	1	1	2	2	3	3	4	4
88	9445	9450	9455	9460	9465	9469	9474	9479	9484	9489	0	1	1	2	2	3	3	4	4
89	9494	9499	9504	9509	9513	9518	9523	9528	9533	9538	0	1	1	2	2	3	3	4	4
90	9542	9547	9552	9557	9562	9566	9571	9576	9581	9586	0	1	1	2	2	3	3	4	4
91	9590	9595	9600	9605	9609	9614	9619	9624	9628	9633	0	1	1	2	2	3	3	4	4
92	9638	9643	9647	9652	9657	9661	9666	9671	9675	9680	0	1	1	2	2	3	3	4	4
93	9685	9689	9694	9699	9703	9708	9713	9717	9722	9727	0	1	1	2	2	3	3	4	4
94	9731	9736	9741	9745	9750	9754	9759	9763	9768	9773	0	1	1	2	2	3	3	4	4
95	9777	9782	9786	9791	9795	9800	9805	9809	9814	9818	0	1	1	2	2	3	3	4	4
96	9823	9827	9832	9836	9841	9845	9850	9854	9859	9863	0	1	1	2	2	3	3	4	4
97	9868	9872	9877	9881	9886	9890	9894	9899	9903	9908	0	1	1	2	2	3	3	4	4
98	9912	9917	9921	9926	9930	9934	9939	9943	9948	9952	0	1	1	2	2	3	3	4	4
99	9956	9961	9965	9969	9974	9978	9983	9987	9991	9996	0	1	1	2	2	3	3	3	4

INDEX

Absolute expansion of mercury, 218.
humidity, 164.
temperature, 136, 145.
zero, 128.
Acceleration, angular, 78, 85.
linear, 9.
uniform, 9.
Acceleration machine, 11.
Adjustment of cathetometer, 109.
Air, correction in weighing, 173.
table of density of dry, 227.
Alluard, form of hygrometer, 167, 170.
Apparent expansion, 216.
Archimedes' principle, 172.
Aspirator, 160.
Atomic theory, 138.
weights, 144.
August, wet-and-dry bulb hygrometer, 169.
Avogadro, law of, 131, 138, 143.

Balance, analytical, 116.
Mohr's, 175.
simple, 36, 37, 38.
spring, 26.
Barometer, 108.
corrections, table of, 229.
Bernouilli, work on the kinetic theory, 106.
Black, theory of heat, 198.
Blow-pipe, 149.
Boiling temperature, 133, 156.
Boyle's Law, 105.
departures from, 107.
Brake, Prony, 48.
Brunner, the chemical hygrometer, 168.
Bunsen ice calorimeter, 208.
Buoyancy, correction for, 173.

Cailletet, liquefaction of gases 128
Caliper, micrometer, 223.
vernier, 224.
Caloric, 198.
Calorie, definition of, 198, 203.
Calorimeter, ice, 208
steam, 208
Calorimetric methods, 204.
Calorimetry, 198.
Capillary constant, measurement of, 186, 194.
correction, 159.
depressions, table of, 229.
phenomena, 182.
tubes, change of level in, 185.
Carbon dioxide, solid, 128.
Cathetometer, 109.
Center of gravity, 34, 35, 37.
of inertia, 33, 35.
of mass, definition of, 33, 34.
of oscillation, 96.
Centripetal force, 100.
Chappuis, determinations of gas coefficients, 126.
Charles's Law, 122.
Chemical hygrometer, 168.
Circular motion, uniform, 101.
Clausius, development of the kinetic theory, 106.
Coefficient of expansion, definition of, 215.
of solids and liquids, 216.
of gases, 124.
of glass, 133.
table of, 230.
Coefficient of restitution, 58, 59.
of rigidity, definition of, 72.
of elasticity, table of, 226.
Coincidences, method of, 96.

Component of a force, definition of, 23.
Composition and resolution of forces, 21.
Compound pendulum, 95.
Conservation of energy, 45.
Constant proportions, law of, 138.
Constant of radiation, 204.
Constant-volume thermometer, 125.
Contact angle, 187.
Contractility of a surface film, 192.
Cooling, curves of, 210.
 by evaporation, 155.
 law of, 205.
 method of, 206.
Correction of a mercury thermometer, 135.
Curve of saturated vapor, 152.

Dalton, development of the atomic theory, 138.
 law of multiple proportions, 140.
Daniell, dew-point hygrometer, 166.
Density, of air, 114.
 of dry air, table of, 227.
 of carbon dioxide, 145.
 of gases and vapors, 138.
 of gases, table of, 144.
 of a liquid, 174.
 of mercury, table of, 227.
 of a solid, 174.
 of solids, liquids, and gases, table of, 226.
 of water, table of, 227.
Dewar, liquefaction of gases, 128.
Dew-point, 165.
Dew-point hygrometer, 166, 167.
Dimensional formulae, 84, 85.
Disk, inertia, 82.
Donny, superheating of water, 157.
Double weighing, 39, 173.
Dufour, superheating of water, 157.
Dulong and Petit, law of, 202.
 method for expansion of mercury, 218.
Dyne, definition of, 17.

Efficiency, definition of, 46.
Elasticity, 65.
 coefficients of, 65.
 table of, 226.
Energy, definition of, 42
 conservation of, 45
 and efficiency, 42
 kinetic, 42, 43.
 molecular, 129
 potential, 44.
 of rotation, 84
 unit of, 42.
Equation of moments, 32, 37.
Equilibrant, 33
Equivalence, principle of, 139.
Evaporation, 155.
Expansion, apparent and absolute, 216
 of mercury, absolute, 218.
 coefficient of, 215
 coefficient of gases, 123, 126, 215.
 coefficients of solids and liquids, 215, 218.
 coefficients, table of, 230.

Faraday, liquefaction of gases, 128.
Field of force, uniform, 34.
Films, theory of thin liquid, 189.
Force, 13, 15
 absolute unit of, 17.
 centripetal, 100.
 composition and resolution of, 21, 22, 23.
 constant of a spring, 90.
 definition of, 17.
 gravitational unit of, 17.
 intermolecular, 182.
 table, 26.
Freezing mixture, 135.

"*g*, determination of, 95.
"*g*," value of, 13
Galileo's definition of force, 15.
Gases, theory of, 106.
Gas thermometers, 123.

Gay-Lussac, law of, 122.
experiments on temperature of water and its vapor, 157.
Glass, coefficient of expansion of, 133.
Gram, definition of, 17.
Gram-molecule, 201.
Gravitation, law of, 20.
Gravity, center of, 34, 35, 37.
Gyration, radius of, 96.

Heat capacity, definition of, 200.
latent, 199.
table of latent, 200.
mechanical equivalent of, 45.
molecular, 201.
table of molecular, 201.
definition of specific, 200.
table of specific, 201, 230.
early theory of, 198.
Heats, specific, variation with temperature, 203.
Henry, measurement of capillary constant, 196.
Hooke's Law, 66, 91.
Horse-power, 51.
Humidity, absolute, 164.
relative, 164.
Hydrogen, liquid, 128.
solid, 128.
thermometer, 127.
Hydrometer, constant-weight, 178.
Hygrometer, absorption, 165.
chemical, 168.
dew-point, 166, 167.
wet-and-dry bulb, 169.
Hygrometry, 164.

Ice calorimeter, 208.
Ice, latent heat of, 199.
Impact, elastic, 61.
inelastic, 54.
laws of, 52, 53.
Inertia, 78.
center of, 33, 35.
moment of, 78, 79.
Inertia disk, 82.
International Bureau of Weights and Measures, method of measuring linear expansion coefficients, 219.

Johonnott, thickness of the black spot, 191.
Joule, equivalent of heat, 45, 46.
development of the kinetic theory, 106.
relation between pressure and molecular energy of a gas, 129.
definition of, 51.

Kinetic energy, 42, 43.
energy in impact, loss of, 53, 59.
energy of rotation, 84.
energy of translation, 84.
theory of gases, 106.
theory of liquids, 152.
theory and temperature, 127.

Laplace, formula and its deduction, 193.
theory of molecular pressure, 183.
normal pressure, 185.
Latent heat, 199.
table of, 200.
Lavoisier and Laplace, linear coefficient, 219.
LeRoy, dew-point hygrometer, 166.
Leslie, wet-and-dry bulb hygrometer, 169.
Lever arm, definition of, 32.
Lever, law of, 31, 32.
optical, 68.
Linear expansion coefficient, 218.
Liquid, density of, 174.
Logarithms, table of, 235.

Manometer, 48, 49.
Mariotte's Law, 105.
Mass, unit of, 17.
gravitational unit of, 17.
definition of center of, 33, 34.
Maxwell, proof of Avogadro's Law, 145.

Mayer, conservation of energy, 45.
Mean free path, 153.
Mechanical equivalent of heat, 45.
Mercury, table of densities of, 227.
Method of coincidences, 96
of cooling, 206.
of mixture, 204, 211.
of oscillation, 117.
Micrometer caliper, 223.
Modulus of torsion, 74.
Modulus, volume, 66.
Young's, 67.
Mohr's balance, 175.
Molecular energy and pressure of a gas, 129.
force, 182.
heat, 201.
heats, table of, 201.
pressure, 183.
pressure, relation to surface tension, 192.
pressure, variation with curvature, 184.
velocity, 106, 130.
weights, 144, 201.
Molecule, 139, 152.
Moment of force, definition of, 32.
of inertia, definition of, 78, 79
of torsion, definition of, 75.
Moments, equation of, 32, 37.
principle of, 30, 31.
Momentum, 15, 43.
conservation of, 52, 53.
Motion, rotary, 81.
simple harmonic, 87.
uniformly accelerated, 9.
uniform circular, 101.
Motor, water, 48.
Multiple proportions, law of, 140.

Neumann, law of, 202.
Newton, interpretation of Third Law, 42.
coefficient of restitution, 58.
conservation of energy, 45.
law of cooling, 205, 211.
law of gravitation, 20.
Newton, laws of motion, 15, 17.
Nitrous oxide, liquid, 128.
Normal pressure, 185.

Olszewski, liquefaction of gases, 128.
Optical lever, 68.
Oscillation, center of, 96.
method of, 117.
Oxygen, liquefaction of, 128.

Parallax, error of, 12.
Parallelogram of forces, 22.
Pascal's Law, 181
Pendulum, compound, 95.
simple, 96.
torsion, 92.
Pictet, liquefaction of gases, 128.
Potential energy, 44.
Pressure coefficient of gases, 125, 126.
exerted by a gas, 129, 130.
law of liquid, 181.
Prony brake, 48.
Pulleys, system of, 47.
Pump, water, 160.

Radian, 68, 75, 79.
Radiation constant, 204.
Radius of gyration, 96.
of influence of molecular force, 183.
Rayleigh, measurement of thickness of oil films, 190.
Reduction of vapor pressures, 159.
Regnault, experiments on Boyle's Law, 107.
device of counterpoise in weighing, 115.
experiments on hygrometers, 166.
coefficient of expansion of mercury, 218.
Relative humidity, 164.
Repulsion theory of gases, 105.
Resolution of forces, 21.
Resultant, definition of, 22.
of forces, 23, 24.

Restitution, coefficient of, 58, 59, 61.
Rigidity, coefficient of, 65, 71, 72.
Rotary motion, 81.
Rowland, experiments on specific heat of water, 203.

Sanctorius, absorption hygrometer, 165.
Saturated vapor, 154.
Saussure, absorption hygrometer, 165.
Scholium to third law of motion, 29.
Sensitiveness of a balance, 118, 177.
Simple harmonic motion, 87, 90.
  pendulum, 96.
Specific heat, definition of, 200.
  table of, 201, 230.
  variation with temperature, 203.
Spring balance, 26.
Steam calorimeter, 208.
  latent heat of, 199.
Stevin, proof of Archimedes' principle, 172.
Substitution, principle of, 141.
Superheating of water, 157.
Surface tension, 156, 190.
  measurement of, 195.
  relation to molecular pressure, 192.
  table of, 230.
  theory of, 190.

Table of capillary depressions of mercury, 229.
  of coefficients of expansion, 230.
  of densities of dry air, 227.
  of densities of mercury, 227.
  of densities of solids, liquids, and gases, 226.
  of densities and tensions of water vapor, 228.
  of densities of water, 227.
  of elastic coefficients, 226.
  of logarithms, 235.
  of reductions of barometer heights, 229.
  of specific heats, 230.
Table of surface tensions, 230.
  of trigonometric functions, 231.
Temperature, absolute, 136, 145.
  boiling, 156.
  measurement of, 122.
Tension in a cord, 19.
  surface, 156, 190.
  measurement of surface, 195.
  theory of surface, 190.
  table of surface, 230.
  of saturated vapor, 158, 165.
Thermometer, gas, 123.
  corrections, 135, 158.
  corrections, curve of, 159.
  hydrogen, 127.
Torsion, modulus of, 74.
  moment of, 75.
  pendulum, 92.
Trigonometric functions, table of, 231.

Uniform accelerated motion, 9, 10.
  circular motion, 101.
  field of force, 34.

Vapor pressures, reduction of, 159.
  saturated, 154.
  tension, 158.
  tension, table of, 228.
Vaporization, 153.
Variation of specific heats with temperature, 203.
Vector, 100.
Velocity, 21
  angular, 84.
  mean, 10.
  molecular, 106, 130.
Vernier, 224.
  caliper, 225.
Vinci, capillary phenomena, 182.
Violle, specific heat of platinum, 213.
Volume coefficients of solids, 217.
  modulus, 66.

Water density, table of, 227.
  motor, 48.
  pump, 160.

Watt, definition of, 51.
Weighing, method of double, 39, 173.
with false balances, 38.
Weight of body in vacuo, 172.
Weights, atomic and molecular, 144.
Wet-and-dry bulb hygrometer, 169.
Woestyn, law of, 202.
Work, definition of, 29.
Work, principle of, 29.
unit of, 42.
Wroblewski, liquefaction of gases, 128.

Young's modulus, 67.

Zero, absolute, 128.

Made in United States
Troutdale, OR
11/03/2026

56845233R00139